From Kyoto to the Town Hall

From Kyoto to the Town Hall

Making International and National Climate Policy Work at the Local Level

Edited by Lennart J. Lundqvist and Anders Biel

Routledge
Taylor & Francis Group

LONDON AND NEW YORK

First published by 2007 Earthscan

2 Park Square, Milton Park, Abingdon, Oxon OX14 4RN
711 Third Avenue, New York, NY 10017, USA

Routledge is an imprint of the Taylor Francis Group, an informa business

First issued in paperback 2016

ISBN: 978-1-84407-423-5 (hbk)
ISBN: 978-1-138-97474-6 (pbk)

Typeset by MapSet Ltd, Gateshead, UK
Cover design by Jonathan Chapman

A catalogue record for this book is available from the British Library

Library of Congress Cataloging-in-Publication Data
From Kyoto to the town hall : making international and national climate policy work at the local level / edited by Lennart Lundqvist and Anders Biel.
    p. cm.
  ISBN-13: 978-1-84407-423-5 (hardback)
  ISBN-10: 1-84407-423-4 (hardback)
   1. Climatic changes—Government policy. 2. Climatic changes—Environmental aspects. I. Lundqvist, Lennart, 1939- II. Biel, Anders.
  QC981.8.C5F757 2007
  363.738'745—dc22
                                                          2006039449

Contents

List of Figures and Tables

FIGURES

TABLES

List of Contributors

EDITORS

Lennart J. Lundqvist   Department of Political Science, Göteborg
University, Sweden
email: lennart.lundqvist@pol.gu.se

Anders Biel   Department of Psychology, Göteborg University,
Sweden
email: Anders.Biel@psy.gu.se

CHAPTER AUTHORS

Monika Bauhr   Department of Political Science, Göteborg
University, Sweden

Kristina Ek   Economics Unit, Department for Business
Administration and Social Sciences, Luleå
University of Technology, Sweden

Lena Gipperth   Department of Law, Göteborg University, Sweden

Henrik Hammar   National Institute of Economic Research,
Stockholm, Sweden

Sverker C. Jagers   Department of Political Science, Göteborg
University, Sweden

Gabriel Michanek   Jurisprudence Unit, Department for Business
Administration and Social Sciences, Luleå
University of Technology, Sweden

Andreas Nilsson   Department of Psychology, Göteborg University,
Sweden

Maria Pettersson   Jurisprudence Unit, Department for Business
Administration and Social Sciences, Luleå
University of Technology, Sweden

Patrik Söderholm   Economics Unit, Department for Business
Administration and Social Sciences, Luleå
University of Technology, Sweden

Johannes Stripple	Department of Political Science, Lund University, Sweden
Chris von Borgstede	Department of Psychology, Göteborg University, Sweden
Mathias Zannakis	Department of Political Science, Göteborg University, Sweden

Preface

This book is a collection of studies on the challenge of transforming the principles of a national climate strategy – derived, in part, from global commitments – into effective local and individual action and behaviour. The individual chapters all emanate from a cross-disciplinary programme carried out at Göteborg University in Sweden called COPE (Communication, Organisation, Policy Instruments, Efficiency), aiming to research ways of achieving the objective of reduced climate impact. The COPE programme was initiated in 2001, and a planning grant from the Faculty of Social Sciences at Göteborg University provided valuable stimulus for bringing several researchers from different disciplines together. The research outline resulting from this initial cooperative effort was then used to seek a planning grant from the Swedish Environmental Protection Agency (SEPA). A fully developed research proposal was then accepted by SEPA in late 2001, and the programme ran between 2002 and 2005. One project was financed by the Swedish Energy Agency (SEA).

The research reported in this volume is the result of our common work within the COPE programme. We look at the climate change issue from an angle that is quite unique in this research field. COPE started from the challenges posed by climate change to effective multilevel governance, recognizing that this issue causes tensions between the European Union (EU) and its member countries over distribution of burdens among Annex I countries and others (Annex I parties include the industrialized countries that were members of the Organisation for Economic Co-operation and Development (OECD) in 1992, plus countries with economies in transition, including the Russian Federation, the Baltic States, and several Central and Eastern European states). International agreements to which nations have committed themselves put pressure on those nations to change regulatory, economic and organizational measures. As for the national dimension, the encompassing character of climate change policy causes tensions over redistribution of burdens among societal sectors, as well as within and between national and local administrations. Within local sectoral administrations, strategies to combat climate change challenge the value structures, norms and roles held by individuals working there. The imposition of different climate policy measures furthermore challenges deeply rooted values and lifestyles of individual citizens. In the end, effective implementation of national climate policy measures requires individual acceptance and local commitment.

In the programme, we have analysed individual responses and local actions as expressions of the dynamics of system components at the lower level, as well as of the opportunities and constraints imposed by the system dynamics at the higher level. The focus in this final report from the programme focuses on examining efforts to bring about adequate *individual* responses and effective *local-level* cooperation and joint management programmes to implement state policies that are – to a large extent – derived from national commitments made in international negotiations and agreements.

The contributors to this volume represent several academic fields: environmental law, economics, psychology, political theory, policy studies, international politics and sociology. Originally comprising researchers from political science, psychology, economics and law at Göteborg University, additional SEPA grants in 2002 financed the inclusion of researchers from economics, law and sociology at Luleå University of Technology and a political scientist from Lund University specializing in international politics. Their contributions represent original studies emerging from the seven different sub-projects within the COPE programme. The general outlines of the chapters have been discussed at project and reference group meetings, and they have been presented and discussed in close-to-final form in workshops at the Seventh Nordic Environmental Social Science Research (NESS) Conference held in Göteborg in June 2005. The international members of the COPE Reference Group – Professor Susan Baker from the University of Cardiff, Dr Knut Alfsen from Statistics Norway and Dr Frans Coenen from the Centrum voor Schone Technologie en Milieubeleid (CSTM) at Twente University – all provided valuable comments in special seminars throughout the COPE programme.

As coordinators and leaders of the COPE programme, we would like to express our deeply felt gratitude to all our fellow COPE researchers and our international reviewers for their energy, enthusiasm and – above all – stamina in helping us to bring our research to a fine ending. Cross-disciplinary research is rarely easy to get going and keep together; but we dare say that COPE has proven the exception; thanks to all of you for an exciting journey in a fascinating research landscape! Thanks are due to SEPA and SEA for generous grants and for all kinds of support throughout the programme, and to Kerstin Gidsäter for bringing final order to our manuscript.

Lennart J. Lundqvist and Anders Biel
Göteborg
October 2006

List of Acronyms and Abbreviations

AAU	assigned amount unit
AGBM	Ad Hoc Group on the Berlin Mandate
AIJ	Activities Implemented Jointly programme
C	Celsius
CDM	Clean Development Mechanism measure
CHP	combined heat and power
CO_2	carbon dioxide
COP	Conference of the Parties
COPE	Communication, Organisation, Policy Instruments, Efficiency (a Swedish programme for research into ways of achieving the objective of reduced climate impact)
CPR	common pool resource
CSTM	Centrum voor Schone Technologie en Milieubeleid, Twente University
CV	contingent valuation
EC	European Community
ECCP	European Climate Change Programme
EU	European Union
GDP	gross domestic product
GHG	greenhouse gas
INC	Intergovernmental Negotiating Committee
IPCC	Intergovernmental Panel on Climate Change
IPPC	Integrated Pollution Prevention and Control
IR	international relations
JI	Joint Implementation measure
KLIMP	local climate investment programme
km	kilometre
kWh	kilowatt hour
LIP	local investment programme
MW	megawatt
n	total population sample size
NAP	national allocation plan
NEP	New Environmental Paradigm
NESS	Nordic Environmental Social Science Research

NGO	non-governmental organization
NIMBY	not in my backyard
OECD	Organisation for Economic Co-operation and Development
PCF	Prototype Carbon Fund
ppm	parts per million
PPP	polluter pays principle
R&D	research and development
SEA	Swedish Energy Agency
SEPA	Swedish Environmental Protection Agency
SMHI	Swedish Meteorological and Hydrological Institute
SNEA	Swedish National Energy Agency
SOM	Society-Opinion-Media Institute
SPMs	summaries for policy-makers
SWECLIM	Swedish Climate Modelling Programme
tC/year	metric tonne of carbon dioxide per year
TWh	terawatt hour
UK	United Kingdom
UN	United Nations
UNCED	United Nations Conference on Environment and Development
UNFCCC	United Nations Framework Convention on Climate Change
UNEP	United Nations Environment Programme
US	United States
VBN	value–belief–norm
WMO	World Meteorological Organization
WRI	World Resources Institute
WTA	willingness to accept
WTP	willingness to pay
WWF	World Wide Fund for Nature

1

From Kyoto to the Town Hall: Transforming National Strategies into Local and Individual Action

Lennart J. Lundqvist and Anders Biel

CLIMATE CHANGE: A CHALLENGE TO MULTILEVEL DEMOCRATIC GOVERNANCE

Climate is a collective good affected by both natural and human influences. Even if the importance of human influences relative to natural cycles of change is a matter for debate, scientific evidence is gradually making it clear that human-induced climate change impacts have now begun to show in natural ecosystems. Political debate increasingly deals with how to change individual human behaviour and socio-economic processes and activities in order to avoid climate change reaching magnitudes beyond the resilience of ecological systems and the stability of societal systems. At the same time, the durability of historical, present and projected emissions of greenhouse gases into the biosphere makes it clear that 'climate stability' is not a realistic objective for climate policy. A more realistic view of climate politics is one of multilevel action aimed at keeping the impact of human activities on climate variations within limits of ecological, social and economic resilience.

We regard the recent science-based consensual reports that climate change is, to a large extent, caused by human activities that emit greenhouse gases as tenable. Such activities range from air traffic, with a global reach over industrial belts and urban conglomerations, to local small-scale energy use for heating homes and mowing lawns. This means that effective climate strategies inevitably also require action all the way from global to local levels. Since the majority of these activities originate at the local level and involve individual action, however, climate strategies must literally begin 'at home' to 'hit home'.

Measures directed towards individuals to change habits and lifestyles, and pressures on local governments to take action, must gain legitimacy in order to become effective.

The auspices for effective and legitimate multilevel governance to combat climate change are at the core of this book on Sweden's national climate strategy. But how good are they? To begin with, it would seem as if the activities just mentioned – and their consequences – can be analysed along a continuum ranging from concentrated to dispersed. When both origin and consequences are local, this would seem to favour a local handling of the problem and its solutions. Consequences that reach across and beyond local jurisdictions call for decisions and policies at higher administrative levels (see Naustdalslid, 1994).

Simple as this analytical distinction is, however, it becomes less tenable as a principle for policy recommendations when confronted with the realities of climate change. The very commonness of the atmosphere implies that even geographically limited activities causing greenhouse gas (GHG) emissions may eventually have much wider and long-term consequences. Indeed, climate change challenges the traditional allocation of political and administrative authority. Traditional jurisdictions are organized around 'territories and communities'. They have a limited number of hierarchical levels with broad and 'bundled' competences, expected to last for a long time. The character of the climate issue rather points to a need for reorganization:

- around climate change as a 'problem';
- across traditional levels;
- with task-specific missions and competences; and
- with flexibility to allow for change with increasing knowledge of causes and cures.

Viewing climate governance as such a dynamic system of 'spheres of authority' across different scales may better capture the forces that make – and, perhaps, break – climate change policy than do existing hierarchical models (Hooghe and Marks, 2003).

To begin with, international commitments and national actions create dynamics in multilevel governance. Nation states commit themselves to internationally agreed policies and measures. This means that they agree to supranational control and possible sanctions if they do not fulfil their commitment under the Kyoto Protocol or the European Union (EU) Climate Policy, and that they thus relinquish some of their political authority upward. But once democratic nation states are committed to implement globally decided climate strategy objectives and measures, strategies of 'domestication' are called for. National governments must find ways of making lower governmental levels, as well as private firms and individuals within their territories, take appropriate action to heed the nation's commitments. In so doing, democratic governments will have to observe legitimate claims for local and individual self-government (see Plattner, 2002).

The transformation of international commitments into national policy and further into locally implemented measures provides actors with different, sometimes even contradictory, signals concerning appropriate action. Local decision-makers and individual citizens soon find themselves raising questions such as: 'Why should we act, when our contribution/non-action is hardly discernible?' or 'Should we really engage ourselves in actions to combat climate change when non-participants might benefit without contributing time and resources?' Such questions reveal that climate change brings to the fore the basic tenets of social dilemmas. Viewed as a public good, climate challenges individuals or groups to decide whether to contribute or not to that good. Seen as a common pool resource, climate forces individuals or groups to choose whether to harvest as much as they want from a shared resource or whether to limit their use of the resource. Both types of dilemmas are characterized by free access to the resource and by the fact that cooperative behaviour is voluntary. An example of relevance here is the choice situation facing urban commuters: 'Should I use my own car or opt for public transportation' (see Messick and Brewer, 1983; Dawes and Messick, 2000).

MULTILEVEL GOVERNANCE AND CLIMATE CHANGE: A DYNAMIC PERSPECTIVE

The dynamics and tensions climate change brings to multilevel governance thus provide a formidable challenge to social science research. But how should these phenomena be meaningfully interpreted and successfully attacked? We argue that this can be done by disaggregating them into interacting processes and structures at different scales. A meaningful interpretation of the responses and actions at any one level must seek to simultaneously capture the driving and constraining forces at lower and higher levels. Individual responses and local actions should be seen as expressions of the dynamics of individual norms and values and system components at the lower level, as well as of the opportunities and constraints imposed by the system dynamics at the next higher level (see Cash and Moser, 2000).

Using Sweden's strategy for *Reduced Climate Impact* as a case in point, the research reported in this volume looks at multilevel governance of climate change from this angle. Linked to international climate negotiations and agreements, Sweden's national objective goes beyond the nation's international commitments. Such nationally imposed measures will substantially influence not only local governments and their developmental aspirations, but also market actors' and individual citizens' freedom of choice concerning mobility, production, consumption and the like. At the same time, individual citizens and groups, as well as local governments, can strongly affect the implementation of climate strategy measures in democratic nations. Local governments enjoy constitutionally guaranteed space for local self-governance, and individuals can make use of their constitutional rights and liberties. Our central research question is:

How can different parts of a national climate strategy such as the Swedish one be communicated, organized and instrumented to interact in such a way as to make climate strategy implementation legitimate and effective?

This and the following chapter provide the ground for our analysis of climate policy implementation as a problem of multilevel governance. The rest of Chapter 1 outlines our analytical framework. A major premise underlying the framework is that the impact of human activities on climate is rooted in three current and major shortcomings:

1 *lack of knowledge* about the relationships among human activities, impacts and negative climate consequences;
2 *lack of motivation* to change activities and behaviour that currently contribute to negative climate impacts, even if there is adequate knowledge broadly diffused across all segments of the population;
3 *lack of adequate organizational and legal structures* for effective measures against negative impacts and effective management of climate as a collective resource, even if there is both adequate knowledge and motivation to make counteraction legitimate.

The rest of this chapter outlines how these three shortcomings create tensions in multilevel governance, and how policy instruments and organizational patterns provide for dynamics affecting the possibilities of overcoming these shortcomings in order to bring future climate variations within limits of ecological, social and economic resilience.

Chapter 2 provides a more detailed description and argumentation to support our choice of Sweden as a critical case for studying the dynamics of multilevel governance. First of all, Sweden's present climate situation and the projected trends in GHG emissions and changes in climate indicate some quite dramatic scenarios. Second, the national objective to reduce national emissions of greenhouse gases by at least 4 per cent, on average, from 1990 levels by the period of 2008 to 2010 goes beyond the reduction demanded by the EU as Sweden's contribution to the implementation of the EU's climate strategy. In a longer-term perspective, Sweden has adopted *Reduced Climate Impact* as one of its 16 intergenerational national environmental objectives. This would require a reduction of at least 50 per cent in current levels of Swedish GHG emissions up to the year 2050. National policy thus puts strong pressure on local governments, business and individual citizens to change climate-affecting habits and behaviour. Third, the political and administrative context in which the national climate strategy is implemented provides for both opportunities and obstacles to effective and legitimate transformation of national objectives into local and individual action. The constitutionally guaranteed powers and administrative organization of local governments make them strong and crucial actors in the implementation process. Historic patterns of cooperation with business and non-governmental organizations (NGOs) on environmental issues are also highly significant in that process.

4

COMMUNICATING SCIENTIFIC INFORMATION IN ORDER TO AFFECT INDIVIDUAL CONSCIOUSNESS AND RESPONSE

If *lack of knowledge* is deemed a primary cause of climate change, *information* and its communication come into focus. The fate of climate strategies is heavily dependent upon citizens' acceptance of scientific information about the causes, further trajectories and consequences of climate change. Given the political difficulties of global climate negotiations, the degree of unanimity and certainty among scientists and policy-makers about the state of the climate and adequate countermeasures is crucial to the success of any climate strategy. However, boundaries between science and policy, as well as between relevant and irrelevant, and useful and useless knowledge are partly *socially* established. Different players act strategically by drawing limitations between knowledge and policy to suit their own interests. What makes some scientific knowledge on climate change more relevant than other knowledge actually depends not so much upon its content as upon the process by which it is developed and validated by the different agents involved (Jasanoff and Wynne, 1998). The acceptance of climate strategy measures, and the behavioural changes sought by that strategy, thus depends upon the organization and communication of knowledge among scientists, policy-makers and the general public.

At the receiving end of knowledge communication, personal experience seems more forceful than indirect evidence (see Fazio and Zanna, 1981). Personal behavioural experience enhances attitude clarity, confidence and certainty, as well as knowledge about the attitude object. Indirect evidence may contribute to attitude formation, but attitudes based on indirect experience translate less easily into action. In addition, potentially negative effects from current behaviour do not materialize immediately, but in the distant future. Uncertainty about the longer-term environmental consequences tends to make individuals optimistic and causes them to downgrade the risk that environmental resources are in danger (Gärling et al, 1998).

The climate issue does exhibit such longer-term uncertainties. The timing and methods of communicating information about causes, trends and the relevance of individual behaviour for climate change thus become crucial for policy legitimacy and effectiveness. The national information campaign, launched in 2003, immediately met with credibility problems. The message about the seriousness and consequences of climate change ran counter to the actual weather conditions during the campaign. As a result, the responsible agencies called off the second- and third-year follow-up campaigns. This again points to the importance of the origins and content of climate change information. Are there ways of communicating climate-related information that enhances individual citizens' trust, not only in the content of the message but also in the institutions and actors engaged in multilevel governance to implement climate policy?

PROMOTING INDIVIDUAL AND LOCAL RESPONSE THROUGH MARKET-BASED AND REGULATORY POLICY INSTRUMENTS

Information is, however, mainly complementary to other measures. When *lack of motivation* is a primary cause of present climate problems, other policy instruments and measures come to the fore. We have already pointed to the character of climate as a collective resource. As such, it puts people in a social dilemma where there is conflict between individual and social motives. When individual motives are upfront, people may argue that 'My contribution to the climate problem is negligible', or 'No one will ever notice that I free ride', or 'I have already done all I could be reasonably expected to do.' Indeed, climate policy implementation here faces individual reactions that seem rational in the context of social dilemmas. Why should the driver leave the car when the effect of such a move on GHG emissions is infinitesimal? And why do so if, and when, almost all others continue to use their cars? To paraphrase Hamlet, such thoughts make 'enterprises of great pith and motion ... lose the name of action'.

Indeed, this structural context and individual reactions show how difficult it may be to implement climate policy instruments that target groups regard as placing them in 'Catch 22' situations. An effective climate policy should thus include policy instruments that give quite precise notions about desired and appropriate behaviour. Both regulatory and market-based instruments provide positive and negative stimuli for changing motivations towards behaviours that contribute to upholding the quality of climate as a collective resource. However, the effectiveness of instruments is crucially dependent upon target group acceptance – that is, the perceived legitimacy of different policy measures. Our analysis of policy instruments and their effects on climate strategy implementation departs from the following five questions:

1   Which are the targeted actors and activities?
2   What motives influence targeted human behaviour?
3   What can the (combined) use of instruments achieve?
4   How do different instruments interact?
5   Do instrument(s) have possible negative consequences?

Climate is a collective good whose continued capability of providing services is important for *all* humans and human activities. Paradoxically, this means that climate change as a political issue has no 'natural' or coherent political constituency, at the same time as the selection and design of policy instruments will somehow touch all societal actors and activities. The legitimacy of climate policy thus hinges on how 'each and everyone' is included in the process of selecting and implementing policy measures. However, even if 'all' are in some sense targets for climate policy, they do not have access to processes where international commitments are made. We also know from empirical studies that formal structures in environmental policy-making do have distributive repercussions. Resourceful interests come further into the corridors of power than less resourceful ones. They thus gain much closer access to actual political decision-making compared to more numerous, dispersed and unorganized consumers (Uhrwing, 2001).

As for policy instruments, we build on what is known about the effectiveness and efficiency of market-based and regulatory environmental policy instruments, as well as about the consequences of different combinations of contextual factors and instruments. This knowledge brings us beyond the old dividing line between regulatory 'command-and-control' and 'economic' policy instruments. Given the different conceptions of what the targets really are (e.g. 'cost-effectiveness' or 'just distribution'), it is necessary to define as succinctly as possible the different criteria applicable to enlightened instrument choice. *Cost-effectiveness* means that if a chosen instrument operates as planned, it will achieve the environmental goals at least cost. *Incentive compatibility* means that the agents involved, particularly the polluters, but also regulators, victims and others, have an incentive to actually provide information and undertake abatement as intended. Distributional *equity* concerns what distribution of costs and benefits is 'fair'. The dynamics and tensions of multilevel governance are seen in terms of a struggle between different groups with different notions of these criteria – a struggle that, in the end, affects the legitimacy of the climate strategy. Incentives for acceptance of an instrument may be scarce, particularly if actors feel they lack information or alternatives for action.

But why should not stakeholders 'always act on the cost-reducing options available to them'? The transport sector is a case in point here. Swedish households have become extremely dependent upon private cars as a major means of transportation. The number of passenger cars rose from 3 million to 4 million between 1980 and 2000 in a population of roughly 9 million. A shift to commuting by collective means of transportation is, of course, dependent upon the availability of that alternative. This varies considerably among regions and municipalities, leaving, in particular, those commuting from sparsely populated areas to cities with little choice but to travel by car. Then comes a climate policy initiative of raising the carbon dioxide (CO_2) tax, with an expectation that people react accordingly. To many people dependent upon their cars for commuting and other purposes, the 'rational' choice of choosing collective means of transportation is simply not available. Nor may people see the compensation gained through lower direct income taxes as sufficient to switch to a new, less GHG-emitting (but perhaps more expensive) vehicle.

Individuals and groups thus differ in their sensitivity towards climate issues. They have different motivations to observe regulations, heed information or react to economic incentives or disincentives. Certain actors are more prone than others to interpret policy measures in terms of their opportunities or consequences for individual gain or loss. This is a primary motive that might work against behavioural change and cooperation for the sake of the collective good. Other actors look more to effects in terms of distributive justice and equality across communities or societies. Furthermore, different instruments may in themselves appeal to non-identical belief systems within the same target group. Hence, we pay specific attention to the interplay between decision-

Table 1.1 *Framework for analysing the legitimacy and effectiveness of policy instruments*

Factor	Policy instrument
Organizational and legal structures	Legal (regulatory) instruments Voluntary agreements
Knowledge	Information
Motivation	Subsidies Taxation; tradable emission permits

making and instrument choice, on the one hand, and implementation and use of instruments, on the other.

Several factors can be expected to contribute to behavioural changes and cooperation. Individuals with good knowledge and high consciousness about climate change and its effects may be more prone to change their behaviour in a climate-friendly direction. The same is true when people show a readiness to observe social norms about what is morally right. A belief in one's capacity to contribute to the solution of climate-related problems, and a clear view of the effectiveness of such a contribution, may also contribute to behavioural change. All of these factors may, in turn, be important for what types of policy instruments to select and use. Our common framework for analysing the legitimacy and effectiveness of policy instruments can be summarized in Table 1.1.

ORGANIZING FOR COORDINATED CLIMATE ACTION: A MULTILEVEL DILEMMA

Even if there is both knowledge and motivation, climate policy may still stumble because there is a *lack of effective organizational structures*. This leads to a search for organizational alternatives that ease the transformation of knowledge and motivation into behaviour and activities more beneficial to the climate. Sweden's emphasis on local-level action to reduce climate impact clearly brings to the fore the tension between the two principal models for organizing multi-level climate governance – that is, the area-specific, multipurpose, hierarchical and stable model versus the task-specific, cross-level and flexible model. The constitutionally strong and independent local governments in Sweden correspond to the first model, with different policy sectors and administrations having different goals and constituencies. National efforts to implement the climate strategy thus encounter a local power structure and organization, where both politicians and administrators may question the strategy in terms of legitimacy.

When charged with implementing nationally imposed climate policy instruments based on longer-term concerns for ecologically sustainable development, local policy-makers may, indeed, perceive this as placing them in a dilemma. This is because local governments encounter different logics of action, all affecting the possibilities of implementing national climate policy effectively through a strategy of coordination and cooperation (see Lundqvist, 1998). When local politicians take action to *mobilize resources*, local government's monopoly of physical planning clearly invites them to adopt an exclusively *intra-municipal* perspective to maximize the developmental potential and attractiveness of the municipality. Climate policy instruments such as national support for municipal climate investment programmes create a dynamic of individual local government activity rather than inter-municipal cooperation, provided that the individual municipality deems itself capable of developing an application that will result in a state grant.

A different logic applies when local governments try to *use available resources efficiently*. There are strong incentives for local governments to engage in *inter-municipal* cooperation to gain economics of scale in, for example, large infrastructure investments. Examples that may create a dynamic of cooperative rather than individual local action concern inter-municipal cooperation on waste incineration and the mandatory regional–municipal cooperation on collective transport systems.

More than anything else, climate policy concerns how to *promote resource sustainability*. But in dealing with something as spatially diffuse as climate change, municipal interdependence is not easily distinguishable. Cooperation among local governments might not be forthcoming. One might expect a dynamic of policy implementation that is different among municipalities because of the changing visibility and perceived necessity of climate change and measures to abate it.

In addition, climate change cuts across sectoral or political/administrative borders that often serve to promote and uphold activities which actually contribute to climate change. Local organizational 'cultures' have developed within long-established sectoral administrations – planning, infrastructure development, energy and transport, to name the most obvious. Officials within such administrations might question the legitimacy of climate policy as it comes into conflict with deeply entrenched organizational and professional values and norms. They will do so in different ways, with some showing more and others less willingness to embrace the tasks of climate policy implementation. Reactions in different parts of local government will thus be crucial for the possibilities of effective cross-sectoral local implementation of the national climate strategy.

Such contextual influences on individual decision-makers and administrators may, however, be counteracted by 'guild'-like professional norms among professional groups within or across administrations. Where you 'sit' (i.e. in which administrative unit you are placed) and what you 'do' (i.e. the actual tasks you are performing on the basis of your profession) can thus be expected to frame and form local actors' views of climate change as a problem, as well as of what constitutes appropriate 'solutions'. What is particularly interesting here is the level of acceptance for cross-sectoral, task-specific local cooperation, and what this indicates for the local fate of the national climate strategy.

EFFECTIVE IMPLEMENTATION THROUGH MULTILEVEL INSTRUMENTAL HARMONIZATION

It is not just local political will and administrative and professional culture that make concerted action towards climate change difficult. Decision-makers at varying levels and in different sectors are entangled in a web of regulations and procedures that limit their possibilities of reorganizing and redirecting their activities as new problems occur.

Take the local councillors making decisions to promote a local collective good – for example, a new industrial mall or a new housing estate. They do so in accordance with laws providing for municipal governments' sovereignty in 'local' affairs at the same time as they consciously observe regulations in planning and environmental law. Still, such local decisions may generate new or increased emissions of GHGs with consequences far beyond the municipal borders. Constituting a negative externality, these decisions run counter to national or even global strategies for managing climate as a sustainable collective good. Seen from the local perspective, however, the climate strategies and policy instruments imposed from higher levels seem to counteract local authority to promote social and economic welfare even when consequences within the scope of the local decision-making unit have been taken into account.

These tensions between different levels in the traditional hierarchical system of governance make problems for effective long-term management of climate as a collective resource. The climate issue makes these tensions even more pronounced since it spans all crucial sectoral administrations and all levels of governance. Effective implementation of a national climate strategy based on premises set by international commitments thus necessitates coordination and harmonization of the instruments used in the strategy.

CLIMATE POLICY AND MULTILEVEL GOVERNANCE: A SWEDISH ANGLE TO A GENERAL PERSPECTIVE

The research reported here views the overarching *political* objective of climate policy as aimed at limiting the climate impact of human activities and keeping climate variations within boundaries of ecological, social and economic resilience. To achieve this, different types of policy instruments – legal, market based, informative and organizational – are used, and they have different ecological, economic and social effects. Decision-making on climate strategies and implementation policy measures is distinctly multi-governance in character. Many different actors and organizational levels are involved, from international bodies all the way down to the individual citizen/consumer, all with different views and links to the political context and to other actors.

Such dynamic tensions are, as we have stated earlier, general problems that occur when democratic nations implement globally agreed commitments. Our general perspective contains the following features. First of all, we see climate change and climate policy as creating dynamic tensions in multilevel governance. Second, we see these dynamic tensions as caused by the search for solutions to *social dilemmas* surrounding the management of climate as a common pool resource. Third, our conceptual and theoretical treatment of actors, processes, instruments and their interrelations in multilevel governance is in line with mainstream social science.

The Swedish climate strategy is, in our view, a case in point for analysing the dynamic tensions between the individual and the collective, local and national, and national and global, as well as among different sectors in society. This has to do with the ambitions of the Swedish strategy, as well as with the country's structure of government and history of governance (Cabinet Bill, 2001). We think here of the constitutionally strong local governments and the key role ascribed to them in climate policy. We think further of the strong traditions of governance for consensus linked to the development of the Swedish welfare state.

Since we develop our arguments for choosing Sweden in Chapter 2, we provide here a few examples to motivate this choice. Take the dynamics set in motion by Sweden's opting for a 4 per cent reduction of Swedish GHG emissions rather than going along with the 4 per cent increase allowed by the EU climate programme. Affected interests in the Swedish transportation and energy sectors may question its legitimacy when compared with the burdens put on their colleagues in other EU member states.

Both the Kyoto Protocol and the European Union regulate the size of the national reductions of GHG emissions. Seen from a somewhat different angle, this becomes an accord concerning each individual nation's GHG 'emission rights'. The emission trading system means that the holders – mostly large industries – can buy and sell these rights on an international market. This further means that the rights holders are beyond the reach of the planners and environmental inspectors trying to implement local climate-related measures. Issues of authority, justice and legitimacy thus come to the fore, making for dynamic repercussions all the way down to the local and individual level.

Swedish local governments may view such internationally agreed and nationally applied measures as a limitation on their 'key role' in planning for local socio-economic development. It might also affect the possibilities of achieving voluntary climate agreements with important sectors. Furthermore, the dynamic relationship between 'stern' economic measures, on the one hand, and the 'voluntary agreement' aspect of policy implementation, on the other, is a most crucial factor in determining the possibilities of individual policy acceptance.

The analysis of the dynamic tensions among individual, local and national actions and reactions to climate change policy implementation are thus the special theme addressed in this book. The contribution we make by using the Swedish case is an analysis of what actually happens when national climate policy measures meet local structures, values and actors in the implementation stage. The approach is multidisciplinary. The compatibility – or lack thereof – between international agreements and national legislation, and the effects this has on policy implementation is analysed from the perspective of environmental law. International politics contributes through analyses of how the development of different global 'centres of authority' in climate politics affects

11

national climate policy, as well as the conditions for national institutionalization of globally agreed norms for climate change abatement.

Individual reactions to stimuli provided by different types of nationally inserted policy instruments are analysed by scholars from the fields of psychology, economics and political science. One line of analysis concerns how individually and organizationally held values and norms affect the acceptance of climate policy measures. Another uses multiple choice/multiple policy option surveys to canvass people's views of economic instruments, including their views of climate policy measures and their effects on the 'just' distribution of burdens. Political science uses experiments to assess the effects on individuals of communicating consensual scientific information on climate change and what this might mean for the effectiveness of information as a policy instrument. Psychologists and political scientists analyse how local governments respond to the need for national efforts to 'steer' climate strategy implementation, as well as for cooperation across political and sectoral boundaries.

REFERENCES

Cabinet Bill 2001/02:55 (2001) 'Sveriges klimatstrategi', Swedish Parliamentary Record, Stockholm

Cash, D. W. and Moser, S. C. (2000) 'Linking global and local scales: Designing dynamic assessment and management processes', *Global Environmental Change – Human Dimensions*, vol 10, pp109–120

Dawes, R. M. and Messick, D. M. (2000) 'Social dilemmas', *International Journal of Psychology*, vol 35, pp111–116

Fazio, R. H. and Zanna, M. P. (1981) 'Direct experience and attitude-behavior consistency', in Berkowitz, L. (ed) *Advances in Experimental Social Psychology*, vol 14, Academic Press, San Diego, CA, pp161–202

Gärling, T., Biel, A. and Gustafsson, M. (1998). 'Different kinds and roles of environmental uncertainty', *Journal of Environmental Psychology*, vol 18, pp75–83

Hooghe, L. and Marks, G. (2003) 'Unraveling the central state, but how? Types of multi-level governance', *American Political Science Review*, vol 97, pp233–243

Jasanoff, S. and Wynne, B. (1998) 'Science and decisionmaking', in Rayner, S. and Malone, E. L. (eds) *Human Choice and Climate Change, Vol 1: The Societal Framework*, Battelle Press, Columbus

Lundqvist, L. J. (1998) 'Local-to-local partnerships among Swedish municipalities: Why and how neighbors join to alleviate resource constraints', in Pierre, J. (ed) *Partnerships in Urban Governance: European and American Experiences*, Macmillan, London

Naustdalslid, J. (1994) 'Miljøproblema, staten og kommunane', in Naustdalslid, J. (ed) *Lokalt miljøvern*, TANO, Oslo

Messick, D. M., and Brewer, M. B. (1983) 'Solving social dilemmas: A review', in Wheeler, L. and Shaver, P. (eds) *Review of Personality and Social Psychology*, vol 4, Sage Publications, Beverly Hills, CA

Plattner, M. F. (2002) 'Globalization and self-government', *Journal of Democracy*, vol 13, pp54–67

Uhrwing, M. (2001) *Tillträde till maktens rum: Om intresseorganisationer och miljöpolitiskt beslutsfattande [Access to the Rooms of Power: Interest Organizations and Decision-Making in Environmental Politics]*, Gidlunds, Hedemora

2

Coping with Climate Change: Sweden's Climate Strategy as a Case in Point

Lennart J. Lundqvist and Anders Biel

To underline our argument why Sweden is a case in point for analysing the tensions in multilevel governance brought about by climate change, first, this chapter provides a picture of Sweden's present climate situation, including the trends in greenhouse gas emissions (GHGs) and some projections of future consequences. Second, we outline the major tenets of the national climate strategy, including Sweden's international commitments under the Kyoto Protocol and the European Union (EU) climate policy. Third, in order to illustrate the relevance of the Swedish case, we discuss the political and administrative context of the climate policy. Fourth, we further elaborate upon the relevance and implications of Sweden's climate policy as a critical case for studying more general issues of multilevel governance dynamics and conflicts.

SWEDEN'S CLIMATE: DETERMINANTS AND CHARACTER

Several features make for variation in Sweden's climate. First, the country is quite far north (between 55° and 69° N), which means that its territory stretches for more than 1500km from the southern shores of the Baltic to the northern mountain borders of Norway and Finland. This high latitude provides for long hours of daylight in summer and short days in winter. North of the Arctic Circle (66° N) this means midnight sun in midsummer and Arctic twilight in midwinter. Second, Sweden is under the west wind belt where low pressure systems move with south-westerly winds along the polar front that separates warm air from cold air.

The proximity to the Atlantic Ocean and the predominant winds provide for a milder climate that the northern latitudes would indicate. This is particularly the case for southern and south-western Sweden, while the mountain ridge along the western border to Norway provides a hindrance for the milder and wetter Atlantic winds. Southern Sweden is part of the warm temperate zone with deciduous forests. The south-west, from Gothenburg to Malmo, experiences quite mild winter temperatures, with only short periods of snow and coastal sea areas only rarely freezing. The rest of Sweden has a cold temperate climate with snowy winters and coniferous forests as the dominant vegetation type. The enclosed waters of the Baltic Sea often freeze in winter, especially further north, which means that the east coast of Sweden is much colder.

The low pressure systems of the west wind belt provide for much precipitation throughout the year. Precipitation is richer in the western parts, especially in higher mountains, and is generally higher in summer than winter. North of a line through the great lakes of Vänern and Mälaren, much of the winter precipitation is snow. Winters become progressively longer and colder towards the north of the country, and the average number of days with a mean temperature below the freezing point is much higher in the far north than in the south.

CLIMATE CHANGE IN SWEDEN: TRENDS, PROJECTIONS AND CAUSES

According to the Swedish Meteorological and Hydrological Institute (SMHI), there was a 1° Celsius (C) increase in average temperatures during 1991 to 2004 compared to the 'norm period' of 1961 to 1990. There is a more pronounced increase in the middle parts of the country. It furthermore seems as if the increase is higher during winter, amounting to almost 2°C in the middle and northern parts of Sweden, but with no observable changes in the south-western areas of the country (SMHI, 2005a; see also MSD, 2005).

Precipitation has also increased in the 1991 to 2004 period across the country. In some parts, particularly in the south and in the northern mountainous regions, the increase is above 10 per cent. The SMHI points out that the trends in temperature and precipitation seem to have consequences for both flora and fauna. The tree limit is climbing upward in the mountains and the glaciers are diminishing. The snow season is markedly shorter in south Sweden. Migratory birds are returning earlier in spring, and animal species preferring a warm climate are appearing further north than before (SMHI, 2005b).

The Swedish Climate Modelling Programme (SWECLIM) produced several regional climate scenarios for longer-term future changes in Sweden's climate. Based on assumptions about future carbon dioxide (CO_2) levels in the atmosphere, SWECLIM calculates that for every 30 years from 1990 to 2100, mean temperatures will rise by 1°C – that is, by nearly 4°C in one century. Precipitation might increase by up to 30 per cent during the same period, with some pronounced regional variations. The northern part of Sweden might experience increases of 30 per cent and more, while mean increases for the southern part stay at 20 per cent or even less in some areas. SWECLIM's scenarios project an annual mean temperature rise in Central Europe of 6°C over the same period, possibly causing extreme weather conditions (SMHI, 2006).

One recent study argues that there is only a very small probability that these recent increases are occurring solely because of natural variability. The authors estimate that about half of the warming and about 30 per cent of the increase in precipitation is due to 'anthropogenic forcing'. Making a probabilistic forecast for the Swedish climate in the years 2001 to 2010, the study suggested a 95 and 87 per cent probability, respectively, of warmer and wetter annual means compared to the average for the norm period of 1961 to 1990 (Räisänen and Alexandersson, 2003).

What, then, are the dominant anthropogenic forces at work here? Reports from the SWECLIM research programme emphasize that the major driving force for these projected changes is anthropogenic. The distribution of greenhouse gas emissions indicates dominance for the supply and consumption of energy in industry, housing and the service sector. Together, they account for almost half of total emissions, but their share is declining. From Sweden's 2004 *National Inventory Report* to the United Nations Framework Convention on Climate Change (UNFCCC), one finds that emissions of greenhouse gases in Sweden, calculated as CO_2 equivalents, amounted to 69.6 million tonnes in 2002, which meant that emissions actually came down compared to the 1990 figure of 72.1 million tonnes. The reductions occurred in the following sectors: waste, agriculture and in the housing and service sector's production of energy. Absorption in forests and fields was 30.3 million tonnes in 2002, a small increase from 1990 (SEPA, 2004).

The strongest increase during the 1990s comes from the transport sector, whose share of total GHG emissions reached 30 per cent in 2002. This is due to the increases in:

- passenger cars in actual use from 3 million in 1980 to 4.1 million in 2004;
- trucks and lorries from 300,000 in 1990 to 440,000 in 2000; and
- vehicle kilometres from 5.65 billion kilometres in 1999 to nearly 6 billion kilometres in 2002 (SIKA, 2003).

Agriculture's share was 12 to 13 per cent after falling during the 1990s. Industrial processes accounted for almost 8 per cent of total emissions, holding the same share throughout the period. The share from landfills of waste was just under 3 per cent in 2002.

Total GHG emissions in Sweden are projected to stay below the 1990 baseline up to the year 2010. The major sectors seem to have different trajectories. The energy sector decreases somewhat, while industrial processes are expected to emit more GHGs in 2010 compared to 1990. Agriculture and waste will continue their downward trend. All in all, the prognosis for 2010 is slightly lower than the 1990 baseline. However, GHG emissions are then expected to rise again, to reach levels clearly above the 1990 'target' line (see the objectives set out in Sweden's climate policy in the following section on 'Sweden's climate strategy: A brief outline of objectives and measures').

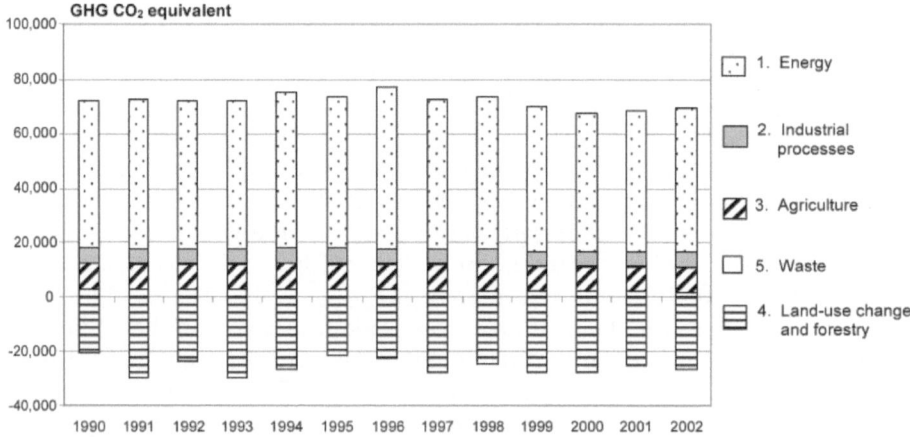

Notes: 'Solvents and other product use' represents too tiny an element to show up in the figure (see Table 2.1).
Figures are corrected for temperature and rainfall to reflect a normal year.

Source: SEPA (2004); MSD (2005)

Figure 2.1 *Total emissions of all greenhouse gases calculated as equivalents from the different sectors: Sweden, 1990–2002*

GHG emissions from the energy sector are expected to show future increases to reach levels almost 14 per cent above the 1990 baseline figure. According to the 2004 National Inventory, the energy sector includes electricity and heating production combustion in the industrial, domestic and service sectors, as well as in refineries and transport. These activities stood for more that 90 per cent of all Swedish CO_2 emissions in 2001. Emissions from industrial processes are on the rise and are expected to be 10 per cent higher in 2020. As for transport, the rate of increase will be close to that of the last decade, while the picture for energy production indicates a break with earlier trends.

Table 2.1 *Historical and projected greenhouse gas emissions by sector: Sweden, 1990–2020*

Sector, 1000 tonnes of CO_2 equivalents	1990	2010	2020	Percentage change, 1990–2020
Energy excluding transport	34.8	33.2	36.2	+4%
Industrial processes	5.7	6.1	6.4	+12%
Transport	18.9	22.6	25.0	+32%
Agriculture	9.6	8.1	8.1	−16%
Waste	2.8	1.2	0.7	−76%
Solvents	0.4	0.2	0.2	−41%
Total emissions	72.2	71.5	76.6	+6%

Source: MSD (2005); but see also SEPA/SNEA (2004)

SWEDEN'S CLIMATE STRATEGY:
A BRIEF OUTLINE OF OBJECTIVES AND MEASURES

The Climate Bill approved by the Swedish *Riksdag* in 2002 recognized that in order to 'meet early on the challenges posed by climate change' Sweden must 'already now sharpen the climate policy' beyond the commitment in the EU distribution of national commitments (Cabinet Bill 2001/02:55, p36). The bill explicitly states that achieving the national objective stated in Sweden's *Reduced Climate Impact* is crucially dependent upon international cooperation and active climate strategies in other countries. Referring to Article 3.13 in the Kyoto Protocol, the bill states that Sweden will stand by its right to save any difference achieved between national objectives and international commitments as a 'basic premise for the Cabinet's proposed national environmental quality objective for climate'. Furthermore, Sweden will reject any demand for international solidarity, implying that it should give some of its 'commitment period reserve' to other member states not fulfilling their assignment under the Kyoto agreement.

Based on this assessment, the government and the Swedish Parliament set as the policy objective to lower GHG emissions to become, on average, 4 per cent less in 2008 to 2012 than in 1990 (measured as CO_2 equivalents). The national objective thus goes far beyond the allowance for a Swedish *increase* of 4 per cent in the EU preliminary distribution of national assignments for GHG reductions. Furthermore, Sweden should reach this objective without compensation for the uptake by carbon sinks, and without using flexible mechanisms. In the longer run, the six GHGs (measured as CO_2 equivalents) should become stable at a level lower than 550 parts per million (ppm) in the atmosphere. By 2050, annual Swedish GHG emissions thus measured should be lower than 4.5 tonnes per capita, and then be brought down further. One should note that the average per capita emissions of GHGs in 2003 was 7.9 tonnes.

To implement the strategy and to reach the objectives, the Climate Bill calls for an 'active and cost-effective' climate policy. There are two key concepts particularly worth mentioning here. One is 'integration': the objective of reduced climate impact is to be reached through a policy that 'is integrated throughout Swedish society'. The responsibility is allocated on 'each and everyone', from central agencies to local governments, business firms, NGOs and individual citizens. This 'integration' of the bill's broad range of policy instruments throughout Swedish society thus provides a new and dynamic context of choice in all walks of life. The other key concept is 'cost-effectiveness': the choice and implementation of climate policy measures must pay due attention to the consequences for 'Swedish industry and its competitiveness'. Table 2.2 provides an outline of the major instruments to be applied in the Swedish climate strategy.

These instruments provide an interesting mix of 'carrots, sticks and sermons'. Notable among the *carrots* is state economic support to local climate investment programmes (KLIMPs). Municipalities, companies and others can apply and compete for grants to finance measures reducing GHG emissions

(SFS, 2003). State support is estimated to have been 900 million Swedish kronor (€100 million, or £66 million) for the period of 2002 to 2004, with a projected additional 200 million Swedish kronor per year for 2005 and 2006. Producers of electric energy from renewable sources get marketable green electricity certificates, and all users of electricity are required to buy a certain amount of certificates in relation to annual consumption. If a user does not buy the required amount of certificates – that is, reaches the mandatory quota – he or she must pay a fee higher than the cost of buying the mandatory quota of certificates. It is thus economically advantageous to buy certificates, which is expected to create a satisfactory demand. The income goes to the producer to cover the extra costs of producing electricity from renewable sources. The trade in certificates creates a market price, and producers who can lower their costs of production may reap a profit. In this way, the certificate system is expected to increase the share of renewable electricity in Swedish energy consumption (Cabinet Bill 2002/03:40). Several measures in the transport policy sector are also of this character, including a tax relief for environmental cars and bio-fuels. To motivate actors to select climate-friendly vehicles, the 2004 Spring Economy Bill proposed an extension of that tax relief until 2008 for cars classified as a taxable benefit when run on environment-friendly fuels, or using environment-friendly technology deemed most favourable to the environment and least detrimental to the climate (see MSD, 2005).

The *sticks* consist primarily of changes in the tax system. The green tax shift set in motion during the 1990s means increasing taxes on energy and fuels in the order of about 30,000 million Swedish kronor (€3300 million, or £2200 million) between 2001 and 2010, while lowering income taxes accordingly. Over the first two years of the tax shift programme, the CO_2 tax was increased by as much as 40 per cent. By spring 2003, the tax had already absorbed about 8000 million Swedish kronor (€900 million, or £600 million). Budget negotiations between the governing Social Democrats and the support parties (the Greens and the Leftists) in September 2003 led to a further tax shift involving 2000 million Swedish kronor (€220 million, or £150 million) in increased energy, fuel and CO_2 taxes, linked to lower income taxes and other tax relief.

Sticks are, of course, also regulatory. There are climate-relevant regulations in the 1998 Environmental Code, the Planning and Building Act and the Municipal Energy Planning Act. Several commissions have investigated the changes necessary to adapt these laws to Sweden's climate objectives and to the obligations and mechanisms that Sweden has agreed to in the international climate negotiation process. The interplay between internationally agreed mechanisms and national regulations provides for interesting dynamics. With the introduction of the EU emissions trading system, the Swedish Parliament altered Sweden's Environmental Code. Since the EU scheme for trading of emissions allowances took force on 1 January 2005, it is no longer permitted for Swedish environmental courts and administrations to prescribe emission limits for CO_2 and to limit the use of fossil fuels for plants needing construction and emission permits under the Code.

Sermons were expected to have an important place in the climate strategy. The 2002 bill proposed a three-year information campaign to increase citizen

Table 2.2 *Major instruments of Sweden's climate policy, responsible levels of government and actors involved in climate governance*

Type of instrument	Specific character of instrument	Formally responsible levels and units of government	Actors/sectors pointed out as crucial to climate governance
Information	Three-year nationwide information campaign	Central: Swedish Environmental Protection Agency (SEPA)	SEPA; local governments; environmental NGOs
Economic	State support to local climate investment programmes	Central: state budget money to SEPA	SEPA; local governments; local business; NGOs
	'Green' tax shift (lower income taxes and higher emission taxes, including CO_2)	Central: Parliament and Cabinet	Taxpayers
	Tax relief for vehicles that run on environment-friendly fuels or use environment-friendly technology	Central: Parliament and Cabinet; Swedish National Tax Board; SEPA	Transport companies; local governments; corporate and individual owners of motor vehicles
	Green marketable certificates with (mandatory) quotas of renewable electricity in power production	Central: Parliament and Cabinet; Swedish National Energy Agency (SNEA); *Svenska kraftnät* (a public utility)	Energy producers; local governments; energy consumers
Regulatory	Measures to ease transfer from non-renewable to renewable sources for energy production, including wind power	Central: Parliament and Cabinet; SNEA; SEPA	SEPA/SNEA; environmental courts; regional administrations; local governments; energy companies
	Changes in environmental and planning legislation to reflect commitment to new climate objectives	Central: Parliament and Cabinet	Central agencies; local governments; actors in such sectors as energy, transport, planning and infrastructure development
	Commitment to implement the flexible mechanisms of the Kyoto Protocol	Central: Parliament and Cabinet	National government (ministries and certain agencies); international actors (EU and Kyoto Protocol process); Swedish industry and business
Other	Voluntary sectoral agreements applying the 'sectoral responsibility' mechanism	Central government (responsible sectoral agencies)	Central agencies; sectoral branch organizations for energy, transport, planning and infrastructure development

Source: MSD (2005)

knowledge and consciousness about the causes and consequences of climate change, as well as about how these can be mitigated and averted. The campaign was to be carried out in cooperation with the Swedish Environmental Protection Agency (SEPA), local governments and NGOs, with a budget of 90 million Swedish kronor (€10 million, or £6.6 million) over the three-year period of 2002 to 2004. The effectiveness of this measure was soon cast in doubt, not just because it meant that each Swede would be subjected to information costing just 10 Swedish kronor (€1.10, or £0.70) per capita over that period, but also because the first campaign round was launched when the weather conditions in Sweden seemed to contradict the core of the campaign message. In fact, the last step including local activities was stymied because this national funding was withdrawn for the campaign's third and final year. In the Swedish national communication to the UNFCCC, this measure is reported as 'no longer in use' (MSD, 2005).

SWEDEN'S CLIMATE STRATEGY: THE CONTEXT OF IMPLEMENTATION

When the Climate Bill calls for an 'active and cost-effective' climate policy that 'is integrated throughout Swedish society', it follows a traditional path in Swedish policy-making: using less of 'command and control' and more of 'cooperation and consensus' between governments at all levels and private-sector interests. A quote from the English summary of the bill captures this special Swedish flavour:

> *This climate work should be integrated with society's activities, and each and everyone should assume his or her share of responsibility. This applies both to national and local authorities and to business enterprises, NGOs and individual persons. Broad participation by all agents will enhance the possibilities of reduced climate impact. Regulations and market-based instruments should be complemented by different agreements and the dialogue between government and business. The climate work conducted by NGOs should also be recognized and supported by national authorities.*
> (Ministry of Environment, 2003, p20)

The implementation of the climate strategy through the different carrots, sticks and sermons is thus a common responsibility for 'each and every' actor in such sectors as energy, transport and infrastructure development (housing, etc.). While the policy instruments might induce each and everyone to change their plans, attitudes and behaviour in a climate-friendly direction, they are, however, expected to do so first and foremost through a process of discussion, negotiations and agreements. In effect, the climate strategy at once outlines two modes of climate policy implementation. One is a complicated pattern of formal governmental responsibilities. The other is a very intricate system of multilevel governance, expected to be driven by a quest for consensual decision-making (see Table 2.2).

The formal system of govern*ment* concerns authority and responsibility among levels and units from the national down to the local level. In the Swedish context, this means that decisions on principles, objectives, instruments and allocation of authority and responsibility for climate policy are made by the Cabinet and Parliament. A specific Swedish feature is that all Cabinet decisions are taken *in pleno* – that is, the Cabinet is collectively responsible. More operational regulations and guidelines are a matter for sectoral governmental agencies. It should be noted that in Sweden, these agencies are formally independent of the Cabinet ministries. This means that the ministers cannot involve themselves in administrative decisions on policy implementation – for example, on how to allocate state grants to local climate investment programmes. Such decisions are the reserve of the appointed 'responsible' agency, in this case SEPA.

The right-hand side of Table 2.2 indicates that formal central government involvement varies among the different types of policy instruments. Decisions on comprehensive economic instruments, such as the carbon dioxide tax, are in the hands of the Cabinet and Parliament. In other cases, such as tax relief for 'green' vehicles or 'green' electric certificates, several agencies become involved, possibly adding to the complexity of multilevel governance.

Our investigations and analyses focus on climate governance at the local level. This is not the least because the 2002 Climate Bill emphasizes the crucial role of Sweden's local governments to establish consensus on climate policy implementation. According to the central governmental view, the 1997 to 2003 state support to local investment programmes (LIPs) for sustainable development led to important reductions of emissions harmful to the environment and climate (see SEPA/SNEA, 2004). Furthermore, the LIP programme provided a good example of building consensus at the local level on measures to promote sustainable development. The extension and redirection of this support to local climate investment programmes thus seemed a good way of securing cost-effective climate mitigation measures, and to encourage municipalities to cooperate both internally and with other municipalities and the private sector. Why this trust of central government in local governance as a vehicle for policy implementation? Is it in any way correspondent with the socio-economic context and political and administrative capacities of municipal government in Sweden?

Swedish municipalities exhibit wide demographic and socio-economic differences (Ministry of Finance, 2004). The municipalities vary in size of population from 760,000 in the capital, Stockholm, to 2600 in the smallest municipality up north. Over half of the municipalities have less than 15,000 inhabitants. It goes without saying that such differences in size are important for both the severity of causes leading to climate change and the capacity and resources to cope with such a complicated issue. Larger cities such as Stockholm and Gothenburg generate more GHG emissions due to heavy traffic, both on land and in the air, as well as from energy production and consumption in industries and in the housing sectors. But smaller municipalities surrounding large road transport arteries, and/or harbouring large industrial complexes with heavy GHG emissions are also strongly affected by the issue of climate change, more so than are municipalities in rural, less densely populated areas. These differences in socio-economic context imply different prospects for successful

climate policy implementation since they are crucial for the individual municipalities' resources and capacities to attract state support for climate-related
action.

Politically, the Swedish Constitution of 1974 recognizes the rights and
authority of the 290 popularly elected municipal councils to promote the
'common interests' of their municipalities. To do this, they are empowered to
levy income taxes on their citizens. Among the mandatory municipal tasks of
particular relevance to climate policy implementation are infrastructure development, environmental and public health protection, waste management, water
and sewage, and rescue services. Of the voluntary municipal tasks of importance for the climate issue, one should particularly mention energy provision
and street maintenance. Together with the regional county councils, municipal
governments are furthermore responsible for regional and local systems of transportation.

Crucially important to climate policy implementation is the local government's political 'monopoly' on physical planning. This empowers individual
municipalities to seek their own developmental paths within the framework of
national law. The changing regulatory context of natural resource and environmental management leading to the 1999 Environmental Code provides
important conditions for such developmental planning. The master plans
(*översiktsplaner*) – formally to be revised every five years – have become a
rolling, more or less continuous, exercise in which municipalities lay out visions
and plans for the use of land, eco-services and the built environment. Swedish
municipalities were also quite active in Local Agenda 21 activities, particularly
in the latter part of the 1990s. It seems plausible that this combination of
planning power and experiences from work on local sustainable development
would provide a local platform for climate policy implementation (see
Eckerberg, 2001; Joas, 2001).

Administratively, the 1991 Municipal Act provided municipalities with
wider latitude on how to organize their own affairs. Except for a municipal
board of directors (elected from the political majority on the municipal council)
and an election committee, earlier mandatory boards and committees can be
substituted by a political and administrative organization tailored to the needs
and political will of the individual municipality. However, there have to be
identifiable administrative units for such tasks that involve the use of authority
towards citizens and groups – for example, building and housing permits, and
environmental and public health protection.

To sum up, Sweden's local governments do have powers and functions of
vital importance to an effective and legitimate implementation of the climate
strategy through multilevel governance. The planning powers make local governments important actors on climate-related issues in the energy, transport and
housing sectors. Local administrations charged with supervision and controls
under the Environmental Code are key actors. Both planning and Local Agenda
21 processes make local governments apprehensive of the problems and possibilities in building consensus on actions towards sustainable development.
However, major parts of these and other sectors are controlled by national
agencies. National policy instruments provide different signals to local actors in

different sectors. Local consensual action, deemed so crucial to effective and legitimate climate policy implementation, might thus not be as easily obtained as the statements by central government suggest.

CLIMATE POLICY IMPLEMENTATION: TENSIONS IN MULTILEVEL GOVERNANCE

Indeed, the climate strategy objectives and instruments provide for tensions among levels of government and actors in governance. To begin with, the regulatory context and instruments give contradicting signals, leading to problems for consensual local governance. There is, for example, a conflict between climate policy objectives, on the one hand, and provisions for affected interests to participate and to forward views and objections in processes of locating facilities for energy production, on the other. For such a climate-friendly technology as wind power, the participatory provisions enable local opinions to question and even stop the establishment of new wind farms.

While the planning powers of local government may seem strong in a comparative perspective, municipalities opting for a strong local climate-friendly agenda may find themselves in a less than favourable position. Large infrastructure investments and facilities, such as railroads, highways, airports and high-voltage power lines, are the prerogative of national government and central agencies. Municipalities find themselves in a supplicant's role; they may modify central infrastructure plans, but cannot actually stop them. Quite naturally, such tensions across levels of government also impact upon the possibilities of creating local consensus; environmental and business groups will most probably push for quite different alternatives.

Another example concerns the international schemes for trading of emission allowances. The Swedish Environmental Code contains prescriptions to set emission limits for CO_2 and to limit the use of fossil fuels. Companies must work within these limits in order to get permits, and local environmental officers have inspection and control authority to make sure these limits are observed. Since the EU scheme for tradable emission allowances was promulgated, however, environmental courts may not prescribe specific legally binding Swedish emission limits for CO_2, as well as limits to curb the use of fossil fuels for these plants.

This adaptation of national policy to international commitments puts local policy implementers in a contradictory decision-making context. On the one hand, the municipality has the power to plan for the future use of resources within its territory, and harbours the environmental administration charged with controlling that emitting facilities adhere to conditions in operation permits. On the other hand, the adaptation to EU schemes means that the major point sources of GHG emissions in the municipality are beyond reach of local policy-makers, planners and environmental inspectors. Furthermore, companies included in the emission trading system have less incentive to engage in local cooperation to abate climate change. Establishing a local governance scheme built on consensus may, indeed, prove difficult under such circumstances.

The economic, or market-based, climate policy instruments also provide for multilevel tensions. To begin with, the state programme for support to local KLIMPs poses a dilemma to local governments. This is particularly so for the smaller, less resourceful, municipalities. Since the sum of available support is so small, local governments may see it as a waste of time to start a process of trying to establish consensus among local actors on a proposal that has little probability of ever gaining support. Furthermore, the green tax shift to increase taxes on climate-affecting emissions is wholly a matter for central government. Local governments cannot decide on specific local tax shift programmes, thus leaving them with income tax hikes as the only substantial source of income. A local government contemplating income tax hikes to abate climate change locally would stand little chance of getting support among its citizens for such a move.

We have so far assumed a united local government attitude in dealing with climate change and in the efforts to build consensual local governance. One should not, however, forget that the all-encompassing character of the climate issue touches and affects governments and actors across levels and sectors with quite varying interests and agendas. Sectoral agencies may have objectives and agendas not fully compatible with the climate strategy. The multitude of responsibilities for economic development, social welfare and sustainable resource use put on local governments make for tensions and conflicts among administrations over the appropriate response to issues of climate change.

All of this indicates that the contextual framework for local – and individual – implementation of the Swedish national climate strategy of *Reduced Climate Impact* provides actors with different, sometimes even contradictory, signals concerning appropriate action. These differences and contradictions affect the possibilities of achieving an effective and legitimate multilevel governance to implement the climate strategy. What is of particular interest here is whether and how Sweden succeeds in building such governance. Historically, Sweden has been a forerunner in environmental policy, with an elaborate scheme of integrating the ecological aspects of sustainable development within its policies and administrative structure. Furthermore, the recent local investment programmes, as well as the many local initiatives and actions under the umbrella of Local Agenda 21, would seem to bode well for Sweden's climate policy implementation.

Thus, we see Sweden as a critical case for successful implementation of climate strategy and policy instruments. If local governance turns out to be problematic here, what then are the auspices for effective and legitimate climate policy implementation in countries that are less developed in these respects?

REFERENCES

Cabinet Bill 2001/02:55 (2001) 'Sveriges klimatstrategi' ['Sweden's Climate Strategy'], Swedish Parliamentary Record, Stockholm

Cabinet Bill 2002/03:40 (2002) 'Elcertifikat för att främja förnybara energikällor' ['Electricity certificates to promote renewable energy sources'], Swedish Parliamentary Record, Stockholm

Eckerberg, K. (2001) 'Sweden: Problems and prospects at the leading edge of LA21 implementation', in Lafferty, W. M. (ed) *Sustainable Communities in Europe*, Earthscan, London

Joas, M. (2001) 'Democratic and environmental effects of Local Agenda 21: A comparative analysis over time', *Local Environment*, vol 6, pp213–221

Ministry of Environment (2003) *The Swedish Climate Strategy*, Summary of Government Bill 2001/02:55, Ministry of Environment, Stockholm

Ministry of Finance (2004) *Local Government in Sweden – Organisation, Activities and Finance*, Ministry of Finance, Stockholm

MSD (Ministry of Sustainable Development) (2005) *Sweden's Fourth National Communication on Climate Change: Under the United Nations Framework Convention on Climate Change*, Ministry of Sustainable Development, Stockholm, Ds 2005#55

Räisänen, J. and Alexandersson, H. (2003) 'A probabilistic view on recent and near future climate change in Sweden', *Tellus Series A: Dynamic Meteorology and Oceanography*, vol 55, pp113–125

SEPA (Swedish Environmental Protection Agency) (2004) *Sweden's National Inventory Report 2004*, Submitted under the UNFCCC, SEPA, Stockholm

SEPA/SNEA (Swedish National Energy Authority) (2004) *Sveriges klimatstrategi – ett underlag till utvärderingen av det svenska klimatarbetet [Sweden's Climate Strategy – A Base for the Evaluation of Sweden's Climate Work]*, SEPA, Stockholm

SFS (Swedish Code of Statutes) 2003:262 (2003) *Förordning om statliga bidrag till klimatinvesteringsprogram [Ordinance on State Support to Climate Investment Programmes]*, SFS, Stockholm

SIKA (Swedish Institute for Communication Analysis) (2003) *Annual Statistics*, SIKA, www.sika-institute.se/statistik

SMHI (Swedish Meteorological and Hydrological Institute) (2005a) *Översikt av Sveriges Klimat [Sweden's Climate: An Overview]*, SMHI, www.smhi.se

SMHI (2005b) *Årsmedeltemperaturer [Annual Mean Temperatures]*, SMHI, www.smhi.se/sgn0102/n0205/arsmedeltemp.htm#norra

SMHI (2006) 'Nytt regionalt klimatscenario för 1961–2100 från Rossby Centre' ('New regional climate scenario for 1961–2100 from the Rossby Centre'), SMHI, www.smhi.se/sgn0106/if/rc/documents/Transientreg_A4.pdf

3

Explaining Public Trust in Institutions: The Role of Consensual Expert Ideas

Monika Bauhr

INTRODUCTION

Societies today face an impressive increase in exposure to expert information. It has become increasingly common to hear experts on the radio and TV defining what kinds of problems societies face and how they should best deal with these problems. Experts tell people what may or may not be dangerous about nuclear power, why people starve in certain parts of the world, how large a population of a certain species is required to be in order to prevent extinction, and what substances contribute to an increased greenhouse effect.

The most intuitive effects of diffusing expert ideas are probably that they may influence beliefs about causes and effects, and possibly alter political preferences. Earlier studies have found such effects from the global diffusion of expert ideas on environmental issues. Typically, studies of the influence of global diffusion of expert ideas focus on their role in redefining political issues, increasing substantial knowledge, and enhancing support for international cooperative solutions (Haas, 1990, 1992; Andresen et al, 2000). Large international organizations, such as the Intergovernmental Panel on Climate Change (IPCC), or networks of like-minded scientists – so-called 'epistemic communities' – produce ideas enjoying wide agreement among experts (hereafter called *consensual expert ideas*) and channel them into the policy-making process.

This chapter suggests that experts also play another important role in international and national environmental management. I argue that the diffusion of consensual expert ideas influences the level of public trust in institutions, such as governments, industry and environmental organizations. Even if the role of expert ideas in redefining or increasing knowledge about political issues is important, an understanding of their influence on public trust in institutions may well be crucial to understanding the role of experts in the local implementation of international agreements. Public trust in institutions plays an important role for the ability of these institutions to achieve policy objectives and to gain popular legitimacy in dealing with political issues. The level of trust in institutions may, for instance, enhance people's willingness to comply with the decisions of the institution (Scholz, 1998). Few, if any, institutions can base their power solely on deterrence. What people think of institutions and how much they trust them is therefore essential to both the acceptance of measures and to the compliance with rules imposed by institutions.

I develop my argument as follows. After a brief introduction to what I mean by consensual expert ideas, I discuss why such ideas may be able to promote public trust in institutions such as government, industry and environmental organizations. I then show how consensual expert ideas actually influenced the level of trust in institutions in two samples of people from Sweden and Tanzania. The differences between these two contexts are considerable. My assumption is that given the large contextual difference between these two countries, it is reasonable to expect that if consensual expert ideas can influence the level of trust in institutions in such different parts of the world, it may also do so in contexts that are more similar. I then discuss these results in the light of three possible lines of explanation. The first I call the 'mere exposure effect', meaning that the mere exposure of institutions to information based on consensual expert ideas may influence public trust in them. Second, there is the 'shared experience hypothesis', which holds that different institutional experiences are key to understanding the influence of consensual expert ideas on public trust in institutions. Third, the 'issue-framing' hypothesis states that experts' ability to frame issues in non-strategic terms may enhance trust in institutions. The two latter explanations are seen as contrasts to the mere exposure effect, which I argue is the most basic and simple explanation for the influence of consensual expert ideas on public trust in institutions. Finally, I confront the ideas presented in this chapter with their wider normative and democratic implications.

WHAT ARE CONSENSUAL EXPERT IDEAS?

Because the stakes are so high and the system so complex, policy-makers cannot rely on popular interpretations of the evidence or on the views of an individual expert. They need an objective source of the most widely accepted scientific, technical and socio economic information available about climate change. (IPCC, 2003)

Consensual expert ideas are defined as expert ideas that enjoy a high level of agreement among experts in a particular field. Most environmental issues, and perhaps especially global environmental issues, are surrounded by a large amount of uncertainty that makes expert opinion diverge, at times quite dramatically. Climate change is no exception. Here, consensual expert ideas on climate change are understood as the information emanating from the

Intergovernmental Panel on Climate Change, which is the organization assigned with the task of producing consensus out of these diverging views.

The IPCC was created in 1988 under the auspices of the United Nations Environment Programme (UNEP) and the World Meteorological Organization (WMO). The IPCC does not conduct any research of its own. Its official role is to 'assess on a comprehensive, objective, open and transparent basis the scientific, technical and economic information relevant to understand the scientific basis of risk of human climate change, its potential impacts, and options for adaptation and mitigation' (IPCC, 2003). These three areas are allocated to three different working groups. Working Group I assesses the scientific aspects of the climate system and climate change. Working Group II assesses the vulnerability of socio-economic and natural systems to climate change, the negative and positive consequences of climate change, and options for adapting to it. Working Group III assesses options for limiting greenhouse gas emissions and otherwise mitigating climate change. Lead authors write the working group assessment reports and then send each part of the reports on an extensive review process in which a very large number of researchers are involved. Approximately 1000 experts are involved in drafting, revising and finalizing the IPCC reports, and about 2500 participate in the review process (IPCC, 2003).

The IPCC produces several different types of assessments on the state of knowledge on climate change. The best known of these are the IPCC assessment reports appearing about every fifth year since 1990. The most politically relevant part of assessment reports are the summaries for policy-makers (SPMs). First, they are politically relevant because they are explicitly formulated to provide policy-relevant information. Second, they are the part of the assessment reports most widely read by policy-makers and journalists. I have used these SPMs here to represent consensual expert ideas on climate change.

The work of the IPCC highlights the difficulties of producing expert consensus around international environmental problems. Since one of the constitutive properties of scientific work is disagreement rather than consensus, the literature assessed by the IPCC necessarily diverges on important points. This is not least because scientific knowledge about such complex systems as the Earth's climate and the effects of its variations are necessarily dependent upon frequently very uncertain assumptions. For example, it is impossible for climate research to take into account all potential variables that might influence the global climate. Besides, what is the baseline? No one knows what the climate would be like if the Earth was left 'undisturbed' by human impact. Getting a unified scientific stance is also difficult because climate change involves such a large number of different areas and disciplines, from social sciences to solar system physics.

The task of producing consensus is even more daunting when we take in the wider range of involved actors. There is not only a heavy scientific input into the content of these reports, based on peer-reviewed and published scientific/technical literature. The official goal of the IPCC is also to include information from industry literature and traditional practices, provided that such knowledge is 'appropriately documented' (IPCC, 2003). Many traditional practices of possible relevance are, however, not likely to be appropriately documented according to the criteria used by the IPCC.

The reports also contain substantial political input. A 'scientization' of policy also means a politicization of science (Lidskog and Sundqvist, 2002). The reason for this is found in the conditions under which science is produced and in the way that consensus is formed within the IPCC. The ties between scientific information and policy-making depend partly upon the situation in which there is a demand for scientific information.[1] Ironically, the demand for authoritative knowledge is strongest when it is the most difficult to deliver. It is at its highest when decision-makers are uncertain about what will happen, and when their decisions have potentially large economic, political and social effects. The science produced under these uncertain conditions and under pressure caused by high decision stakes has been called 'post-normal' in contrast to 'normal' science – that is, the Kuhnian approach to science where 'everyone' unquestionably accepts a given paradigm (Funtowicz and Ravetz, 1985; Jasanoff and Wynne, 1998). The production of consensual scientific information in a 'post-normal' stage is likely to drive science closer to policy-making (see the discussions of 'pure science' and 'trans-science' by Weinberg, 1972, and 'basic' and 'applied' science by Bush, 1945). Since the stakes are high and knowledge uncertain, there is policy input in the process of producing consensus. Within the IPCC, governments and international organizations appoint the lead authors of the assessment reports, identify key policy-relevant issues for IPCC reports (together with other users of IPCC reports) and influence the review process.

The different stakeholders are perhaps most actively involved in producing the summaries for policy-makers. The SPMs are scrutinized and accepted line by line, and often even word by word in a discussion among scientists and government experts that 'takes the form of a debate (very often straightforward negotiations) between lead authors of the scientific report and government experts' (Skodvin, 2000). The purpose is to reach an agreement between scientists and policy-makers on how scientific findings should be properly expressed. At this stage, scientific findings are often 'exposed to massive attack' and efforts to influence their content also occur on non-scientific grounds not only from governments, but also from industry and environmental non-governmental organizations (NGOs) (Skodvin, 2000).

Experts who produce consensus within the IPCC frequently do not agree on important aspects of climate change. Furthermore, the scientists contributing to these reports are scientists who are interested in politics and pursuing a political agenda, and scientists who are not interested in politics at all (see the 'epistemic community' discussion by Haas, 1990, 1992). All told, the consensus produced by these negotiations is probably the closest one comes to consensual expert ideas on climate change, and the IPCC is probably also the organization enjoying the strongest political and mass media authority and attention (Bauhr, 2005a).

EXPLAINING PUBLIC TRUST IN INSTITUTIONS

What influences public trust in institutions? Explanations have been diverse. Economic recessions (Lipset and Schneider, 1987), government performance (Kumlin, 2002), post-material values (Inglehart, 1990), social capital (Putnam, 1993) and mass media framing (Cappella and Jamieson, 1997) are just a few examples of factors that have been believed to influence public trust in institutions (for more extensive overviews of factors influencing public trust in government, see Klingemann and Fuchs, 1995; Borre and Scarbrough, 1995; Kaase and Newton, 1995).

However, the diffusion of consensual expert ideas is typically not used to explain changes in public trust in institutions. Studies on the influence of experts on public opinion often acknowledge the importance of trust in institutions, but do not explain where such trust comes from. In technical language, the level of trust is seen as interacting with the influence of expert ideas on public opinion, not as being an effect of the diffusion of consensual expert ideas. In the words of Jasanoff and Wynne (1998):

> ... the basic framework for public responses (to science and technology) depends largely upon the experience and perception of the trustworthiness of relevant institutions and social actors, not upon the understanding of technical information framed in ways that implicitly take trust for granted. (see also Renn, 1992; Wynne, 1980, 1992)

In that fashion, 'misperceptions' of expert ideas can be explained not as mere ignorance, but as a historically grounded distrust in institutions. For instance, a historically grounded distrust in industry may explain why studies have concluded that people tend to believe that aerosol cans and industrial pollution cause climate change (Jasanoff and Wynne, 1998). Thus, whether or not experts manage to influence public opinion depends upon the level of trust not only in experts, but also in the institutions that are explicitly or implicitly included in the measures that experts suggest should be taken. This implies that if experts would, for instance, suggest an increase in taxes on carbon dioxide as an effective measure against climate change, support for such a measure depends not only upon the level of trust in experts. It also depends upon the level of trust in the governmental institutions responsible for the collection and spending of these taxes.

These studies say little about exactly which factors generate public trust in institutions. I suggest that trust in institutions should not only be seen as an exogenous factor determining the influence of consensual expert ideas on causal beliefs or support for measures. Experts may also *generate* the trust upon which their influence is so dependent. Why is this so? If one of the most probable effects of diffusing consensual expert ideas is its ability to redefine or increase knowledge about environmental issues, could this increased knowledge or reduced uncertainty explain an enhanced trust in institutions? In a sense, increased knowledge could enhance perceived self-efficacy, and perhaps even life satisfaction and happiness, all of which may generate a more positive view, in general, towards the system around us (see, for example, Kornberg and Clark, 1992). However, previous theories and findings within this area seem to suggest that increased knowledge might have a double-edged influence: it may also *decrease* the level of trust in institutions. The rationale behind such development would be that increased knowledge makes people more independent of these institutions. Therefore, the argument goes, people will become more critical towards their competence. Such a development would be in line with the 'new politics' argument, where enhanced political resources and skills are believed to decrease the level of trust in institutions (Listhaug, 1995). It would also correspond to the idea of materialism and post-materialism, where increasing educational levels are thought to lead to post-material values, which, in turn, make people more independent from authorities (Inglehart, 1997).

Thus, other explanations may provide more plausible reasons why consensual expert ideas may increase the level of public trust in institutions. I suggest three possible reasons or mechanisms why this may be so. Consensual expert ideas:

1 associate institutions with authority and impartiality;
2 make people think of their experiences and perceptions of institutions; and
3 frame environmental issues in non-strategic terms – that is, they focus on the substance rather than on what actors may gain from diffusing information or dealing with particular issues.

First of all, consensual expert ideas may associate institutions with authority and impartiality. This explanation builds on the idea that the perceived authority and impartiality of consensual expert ideas 'spill over' onto the institutions associated with this information. The mere exposure of institutions, such as governments, organizations or industry, in connection with consensual expert ideas would thus make people perceive them as more trustworthy.

Although few believe that science really provides conclusive answers – particularly not about complex processes – it does provide some of the most robust proof of knowledge available at a specific point in time and thereby enjoy a fair amount of authority. Some argue that 'the "scientific outlook" has public authority in most parts of the world, with scientists attaining substantial public (even philosophic) standing in world culture'. They see science as having gained an 'extreme legitimacy' and prestige in modern polity (Meyer and Jepperson, 2000). According to this so-called 'world polity school', scientific rationality has become one of the most dominant worldviews today and deeply affects people's thinking about both political problems and the natural world.[2] Experts enjoy authority among publics and decision-makers alike. One sign of the hegemony of the scientific outlook is the emphasis that decision-makers put on the claim that their actions are guided by the best possible scientific advice (Meyer and Jepperson, 2000). The authority of experts is likely to be further enhanced when they produce consensus. This may make people believe that information from consensus-producing organizations such as the IPCC is more trustworthy than information presented by individual experts (Bauhr, 2005b).

Furthermore, expert ideas are often perceived as more impartial than other sources of information. The science of climate change is not 'pure'; policymakers and other stakeholders influence what parts of scientific evidence should make it to the IPCC reports and the SPMs. They also influence how these findings are formulated, and in what context they are placed. However, the ability of consensual expert information to strike a balance between adaptation to political acceptability and scientific impartiality may make consensual expert ideas gain more authority than information from NGOs, industry representatives or individual governments, all of whom are more clearly embedded in various interests. Most likely, the final text is not burdened by any apparent political bias. It can still derive its authority from the scientific stance of impartiality.

Thus, expert impartiality may enhance the perceived impartiality of other institutions when these are explicitly or implicitly associated with consensual expert ideas. For example, when consensual expert ideas claim that the IPCC was initiated by the United Nations or that suggested measures towards climate change presuppose government interventions, such associations may contribute to generate trust in these institutions.

A second possible reason why consensual expert ideas may influence public trust in institutions is that it may make people think of their experience with these institutions. The process of trust formation is somewhat different here. It is first and foremost people's beliefs or experiences with institutions, and not just the exposure of these institutions in consensual expert ideas, that guide whether or not consensual expert ideas can influence the level of trust in those institutions. It is notable that both the first and the second type of explanations build on the assumption that people's perceptions of the ability of institutions to deal with, or perhaps even solve, current problems determine the level of trust. However, the second explanation sees a stronger role for personal experiences and beliefs about these institutions in explaining levels of trust; it is not only the mere exposure of these institutions in consensual expert ideas that determines their levels of trust.

Shared experiences can be both direct and indirect. They can emanate from views of institutions portrayed in the mass media, as well as from direct experiences of, say, corruption in the services that these institutions provide. Such indirectly and directly shared experiences or memories may influence perceptions of which institutions are competent, and perhaps even which institutions should be responsible for dealing with environmental problems

Third, consensual expert ideas may promote an 'issue framing' as opposed to a 'strategic framing' of the issue of climate change (see Cappella and Jamieson, 1997). This explanation relates to the *type* of information that consensual expert ideas promote and diffuse. Consensual expert ideas frame the problem of climate change in ways that make people believe that the goal is to 'solve social ills, redirect national goals and create a better future for our offspring' (Cappella and Jamieson, 1997), rather than showing how actors use the issue of climate change for their own benefit. Thus, in a sense, the difference is between that of altruism and self-interest. This explanation thus suggests that consensual expert ideas put an altruistic frame on the issue of climate change.[3]

Why would issue framing enhance trust in institutions? It is because the institutions involved appear to be suggesting what is in the public's best interest, rather than partaking in the strategic games and negotiations that governments and other actors are involved with on the issue of climate change. Studies have suggested that strategic news reports depicting politics as a game between strategic actors, rather than focusing on political substance, can undermine citizens' trust in politicians and even their faith in the functioning of the political system (Cappella and Jamieson, 1997). So far, however, there is less empirical evidence suggesting that issue framing with a focus on substance would *enhance* the level of trust in institutions.

Thus, the three explanations or mechanisms suggested – the mere exposure effect, the shared experience effect and the issue-framing effect – give us good reasons to believe that consensual expert ideas may be able to influence public trust in institutions. They do seem to make it plausible that consensual expert ideas can have an effect that moves beyond a mere influence on causal beliefs.

COMPARING INFORMATION DIFFUSION AND ITS EFFECTS: AN EXPERIMENTAL DESIGN

The data for this study derives from experiments conducted on university students in Sweden and Tanzania. I chose these two contexts since I wanted to vary macro-level economic, political and cultural differences that I expected to be of relevance to people's perceptions of institutions. The large differences between Tanzania and Sweden make it difficult for consensual expert ideas to have a similar influence in both contexts. Therefore, if consensual expert ideas do, indeed, have a positive influence on public trust in institutions in these two very different contexts, the chances increase for such similarities to appear in more similar contexts.

The experimental design allows us to study the influence of consensual expert ideas in a controlled setting. As experimental stimulus, I used a short film produced out of a selection of IPCC material. The film was designed to make the content relevant and interesting to participants in an attempt to promote readiness to take measures against climate change. In the experiment, participants were randomly assigned to either watch a short film based on IPCC material and to answer a questionnaire afterwards (the experimental group), or to answer the same questionnaire without watching the film (the control group). The influence of the film on participants' answers was thus obtained by comparing how experimental group and control group participants answered the questionnaire.[4]

My selection of IPCC material was based on explorative interviews and pilot surveys among groups of students at the two universities. These interviews and pilot surveys were used for developing a selection process for the film's material and for phrasing the content of the questionnaire employed in the study in a way that would make sense to participants. I made the content of the film include causal beliefs that diverged from participants' previous causal beliefs about climate change, and beliefs that were relevant to the participants' respective contexts. The film also emphasized an expert framing of the climate change issue, and used it in a persuasive attempt to make people act on the issue.

The film contained consensual expert ideas on the causes and effects of, and possible measures against, climate change. The film's introduction stated that the content of the film was based on the information from the IPCC, which is a large organization consisting of several hundreds scientists from all around the world. It also stated that UNEP and the WMO established the IPCC to assess the available scientific information on climate change. Furthermore, this sequence also pointed out that the IPCC information on climate change is the most widely agreed upon information available today and that it is an important cornerstone in all UN agreements on climate change. The film then provided a general description of the greenhouse effect and which are the most important man-made greenhouse gases. It also described the potential effects of climate change and the effectiveness of different measures in combating it. A frequent theme throughout was the role that scientists play in helping people to understand the problems related to the increased greenhouse effect.

HOW CONSENSUAL EXPERT IDEAS INFLUENCE PUBLIC TRUST IN INSTITUTIONS

At this point, let us turn to exploring the influence of consensual expert ideas on public trust in institutions. The analysis is conducted in two different parts of the world and the type of trust asked for is a trust in institutions' information about climate change. Do consensual expert ideas influence public trust in institutions, and if so, how? Can expert ideas have a similar influence in radically different parts of the world? Table 3.1 shows the influence of consensual expert ideas on trust among experimental and control group participants in Sweden and Tanzania in a number of different types of institutions, including government, international organizations, industry, environmental organizations, and such groups as journalists and scientists.

Table 3.1 shows the mean answer of those participants who did not see the film (control group) compared to those who saw the film (experimental group) in Sweden and Tanzania. A one-tailed t-test evaluates whether the answers of the experimental group were statistically significantly different from the answers of the control group. The scale used for these items runs from 1 to 4, where 1 = high confidence; 2 = some confidence; 3 = not much confidence; and 4 = no confidence at all. The total population sample sizes (n) were as follows:

- experimental group, Sweden: 172 to 176;
- control group, Sweden: 191 to 194;
- experimental group, Tanzania: 182 to 193; and
- control group, Tanzania: 194 to 203.

Table 3.1 *The influence of consensual expert information on trust in information about climate change from different institutions*

Indicator	Sweden			Tanzania		
	Control group mean	Experimental group mean	t-value	Control group mean	Experimental group mean	t-value
Nuclear industry	3.1	2.8	−2.90***	3.1	2.6	−4.08***
United Nations	2.3	2.0	−3.36***	2.0	1.7	−3.02***
Environmental organizations	1.7	1.7	−.51	1.5	1.4	−2.21**
Government	2.4	2.2	−2.67***	2.0	2.0	−.107
Central government authorities	2.5	2.3	−3.10***	2.3	2.2	−1.77
Local government authorities	2.6	2.5	−2.08**	2.2	2.2	.03
Scholars within the field (researchers, etc.)	1.4	1.5	1.26	1.7	1.7	.84
Journalists	2.5	2.5	−.53	2.1	2.0	−.19

Notes: * $p < .10$; ** $p < .05$; *** $p < .01$.

The results show that consensual expert information influences the level of trust in institutions' information about climate change. Consensual expert ideas enhanced the level of trust in institutions in both contexts, despite their differences. In Sweden, it enhanced participants' trust in different governmental institutions (the national government, central authorities and local government), the nuclear industry and the UN. In Tanzania, it enhanced participants' trust in environmental organizations, the nuclear industry and the UN. These results could be compared to the general lack of influence of consensual expert ideas on the level of trust in others, including friends, family and neighbours (Bauhr, 2005a).

Table 3.1 also points to interesting similarities and differences between the influence of consensual expert ideas in Sweden and Tanzania. The level of trust in the UN and the nuclear industry was enhanced in both contexts. It is interesting to note that consensual expert information did not influence the level of trust in scholars, despite the fact that they were presented as crucial producers of the ideas in the film. The original level of public trust in this group was high, but remained largely unaffected. Table 3.1 also shows an important difference in the influence of consensual expert ideas in these two contexts: the level of trust in governmental institutions (national government, central authorities and local government) was enhanced in Sweden, but not in Tanzania. In Tanzania, consensual expert ideas enhanced the level of trust in environmental organizations.

INTERPRETATIONS AND IMPLICATIONS

The results presented in this study tell us little about why consensual expert ideas influence public trust in institutions and what are the mechanisms involved. Nevertheless, these results might be used to generate hypotheses on possible explanations for this influence. Attempting to generate hypotheses, I will use my empirical findings to discuss and reassess the three explanations presented earlier. In doing so, I contrast the hypotheses of 'shared experiences' and 'issue framing' with the 'mere exposure' effect, which is considered the most basic explanation for the influence of consensual expert ideas on public trust in institutions.

An effect of 'shared experiences'?

The similar influence on Swedish and Tanzanian participants' level of trust in the UN and the nuclear industry could be explained by the simple fact that these institutions were clearly part of the information diffused. The information portrayed the UN as being involved in the development of consensual expert ideas. The influence of consensual expert information on the level of trust in the nuclear industry's information on climate change may seem surprising. The mechanism at work here may be that expert information on climate change associates the nuclear industry with authority and impartiality and possibly makes it look, for once, like the environmental good guy. Given the continuous publicity on the environmental drawbacks of nuclear power, climate change seems to put the nuclear industry in a somewhat positive light since the experimental group film showed nuclear power to be an energy source that does not increase the greenhouse effect. This may have provided enough reason to believe that information from these organizations is trustworthy.

However, when analysed more carefully, this 'mere exposure' hypothesis does not seem to provide the full explanation. First of all, the level of trust in scientists and experts – that is, those actors most explicitly mentioned in the information used in the experiments – remained largely unaffected. Explicit attention is thus not enough to enhance the level of trust in institutions. Second, the results show some interesting differences between the Swedish and Tanzanian contexts. Clear national differences were found in the influence of consensual expert ideas on institutional trust. These ideas enhanced the level of trust in governmental institutions in Sweden but not in Tanzania. Here, these ideas affected the trust in environmental organizations.

One plausible explanation may be that people living in these two countries share different experiences about the distribution of institutional responsibilities. The results may thus reflect differences in the perceived role and responsibility of the government and environmental organizations in these two societies. Differences in the institutional capacities of the respective governments can lead citizens to have different expectations of what their governments can and should do. Fifty years of strong government could make Swedish participants feel that climate change is a matter for, and within the capacity of, the government. On the other hand, Tanzanians have no such collective experience of strong government. They might, therefore, perceive that other actors are, and perhaps even should be, responsible for these issues. The important role that NGOs play in the management of environmental issues in Tanzania may be the reason for the increased trust in environmental organizations in this context.

An effect of 'issue framing'?

The previous discussion suggested that 'mere exposure' needs to be complemented with 'shared experience' in order to more fully understand how consensual expert ideas affect public trust in institutions. What, then, about 'issue framing'? Could the ability of experts to focus on the character of the climate change problem, rather than on the strategic issues that climate change poses to actors, be used to explain more fully why consensual expert ideas influence public trust in institutions? The trust in the nuclear industry may be an example of such influence. The industry would most definitely have something to gain from a more stringent climate policy, as long as other energy sources are not trusted to produce the amount of energy desired without jeopardizing the climate. Indeed, consensual expert ideas enhanced the level of public trust in information from the nuclear industry. Was the 'issue framing' of consensual expert ideas strong enough to 'hide' the evident self-interest of the nuclear industry? Such an explanation is, however, difficult to separate from a 'mere exposure' effect. We cannot exclude the possibility that the reason why consensual expert ideas influenced public trust in the nuclear industry is the fact that this institution was mentioned in the information used.

On the basis of the material presented here, we are not able to distinguish the differences between these factors. However, the results discussed under the section on 'An effect of "shared experiences"?' suggest that the effects of 'mere exposure' are not the only reason for the influence of consensual expert ideas on public trust in institutions. Other factors seem to be at play in determining this influence. The issue-framing hypothesis could possibly account for some of the influence of consensual expert ideas on the level of public trust in institutions, and, hence, deserves further scrutiny.

CONCLUSIONS

The diffusion of consensual expert ideas has important effects that move beyond traditional expectations. This chapter shows how consensual expert ideas enhance the level of public trust in institutions. These results have important implications since the diffusion of consensual expert ideas is typically not used to explain changes in institutional trust and since many of these institutions may potentially play a central role in climate change management.

An important challenge for the future is to better understand the causal mechanisms and reasons why consensual expert information influences public trust in institutions. The three possible explanations discussed here raise different normative challenges. If the mere exposure of institutions in consensual expert ideas matters for the public's trust in them, then it is, indeed, important to ask what institutions are referred to in information based on consensual expert ideas. Who decides which institutions are mentioned in IPCC reports or in media campaigns that use consensual expert ideas? Do these institutions really 'deserve' to be trusted on climate issues? If the mere exposure of institutions in consensual expert ideas enhances trust, people, at best, use a 'peripheral cue' to establish what institutions are trustworthy. If institutions are associated with a trustworthy source of information they are, in turn, also perceived to be trustworthy. The risk is that the level of the trust in institutions has very little to do with what institutions actually do (see Petty and Cacioppo, 1986).

If *shared experiences* or shared memories also matter, things may be a bit better. People's experiences matter in terms of what institutions they trust. Thus, they do not trust just any institution that happens to be mentioned or associated with consensual expert ideas. However, even past experiences do not provide a guarantee that institutions live up to these expectations – that is, managing climate change in a legitimate way.

The *issue-framing* explanation presents another challenge. Issue framing could possibly make people believe that institutions work for the common good. In a sense, this may have positive effects since such faith could enhance public will to support measures against climate change. However, issue framing does not provide insights into the important strategic political and economic issues that surround the issue of climate change, nor does it highlight which actors have an interest in more lenient or more stringent climate policy. Such knowledge could facilitate an understanding of why and for what purpose actors try to influence opinions – which, in turn, facilitates a critical assessment of the arguments used. Cynicism towards institutions may not be a good thing; but questioning with a dose of scepticism can be healthy. A critical assessment can prove important for channelling genuine public opinions towards the policy process. This would, ultimately, enhance the possibility of a legitimate democratic management of climate change.

NOTES

1   Science may always be conducted somewhere in the interface between scientific discipline and policy-making. In order for science to produce completely impartial knowledge, it must be conducted in a space removed from its possible political applications (Merton, 1973) and from political influence on research areas. This rarely, or perhaps never, happens. Science is always influenced by the pressures of political and economic interests and by an unpredictable world. But even if scientists may not be able to speak 'truth to power' (Price, 1965), the extent to which science is influenced by politics clearly varies.

2   According to Goldstein and Keohane's (1993) widely cited definition, a worldview is 'conceptions of possibilities' that are 'deeply embedded in the symbolism of culture and deeply affect modes of thought and discourse'. Thus, such scientific worldview provides a way of understanding the world.

3   This idea does not suggest that consensual expert ideas, nor the experts producing them, would, in reality, be any more altruistic than other actors. The issue framing of consensual expert ideas may stem from the experts' way of producing knowledge, their role in international society and the expectations of the public. In the words of John Meyer and colleagues (1997), the scientific view of nature 'asserts the existence of a global and interdependent ecosystem that encompasses human beings and sustains the very possibilities of life'.

4   The same film was used in both contexts, and only the accent of the speaker's voice was adapted: a British accent for the Swedish sample and a Tanzanian accent for the Tanzanian sample. This was an attempt to minimize the perceived 'foreignness' of the film in the two contexts. To make causal inferences more certain, I also took some steps to enhance my understanding and interpretation of the results of this study. In both the Tanzanian and Swedish contexts, I complemented the material with focus groups as an aid to interpreting the results. Furthermore, I conducted a parallel 'placebo' experiment in both contexts. In these experiments, participants were randomly assigned to watch a film that on a more general level dealt with changes in weather without linking this to consensual expert ideas on climate change and on human influence on such climate. This was done in order to separate the effects of consensual expert ideas on climate change from just any ideas. This 'placebo' experiment showed no significant effects on public trust in institutions. I also varied the person conducting the experiment in an attempt to account for a possible experimenter effect. However, no significant differences were found between experiments conducted under different leadership.

REFERENCES

Andresen, S., Skodvin, T., Underdal, A. and Wetterstad, J. (2000) *Science and Politics in International Environmental Regimes*, Manchester University Press, Manchester and New York

Bauhr, M. (2005a) *Our Common Climate: How Consensual Expert Ideas Shape Global Public Opinion*, Göteborg Studies in Politics, Göteborg

Bauhr, M. (2005b) 'Att påverka den internationella miljöpolitiken', in Jagers, S. (ed) *Hållbar utveckling som politik*, Liber, Stockholm

Borre, O. and Scarbrough, E. (1995) *The Scope of Government*, Oxford University Press, Oxford

Bush, V. (1945) *Science, the Endless Frontier*, US Government Printing Office, Washington, DC (reissued 1980, Arno Press, New York)

Cappella, J. and Jamieson, K. (1997) *Spiral of Cynism: The Press and the Public Good*, Oxford University Press, New York and Oxford

Funtowicz, S. and Ravetz, J. (1985) 'Three types of risk assessment: A methodological analysis', in Whipple, C. and Covello, V. (eds) *Risk Analysis in the Private Sector*, Plenum, New York

Goldstein, J. and Keohane, R. O. (1993) *Ideas and Foreign Policy Change*, Cornell University Press, Ithaca, NY

Haas, P. M. (1990) *Saving the Mediterranean: The Politics of International Environmental Cooperation*, Columbia University Press, New York

Haas, P. M. (1992) *Knowledge, Power and International Policy Coordination*, University of South Carolina Press, Columbia

Inglehart, R. (1990) *Culture Shifts in Advanced Industrial Societies*, Princeton University Press, Princeton, NJ

Inglehart, R. (1997) *Modernization and Postmodernization: Cultural, Economic, and Political Change in 43 Societies*, Princeton University Press, Princeton, NJ

IPCC (Intergovernmental Panel on Climate Change) (2003) *Introduction to the Intergovernmental Panel on Climate Change (IPCC)*, IPCC, Geneva

Jasanoff, S. and Wynne, B. (1998) 'Science and decisionmaking', in Rayner, S. and L. Malone, E. L. (eds) *Human Choice and Climate Change, vol 1*, Battelle Press, Columbus, Ohio

Kaase, M. and Newton, K. (1995) *Beliefs in Government*, Oxford University Press, Oxford

Klingemann, H. D. and Fuchs, D. (1995) *Citizens and the State*, Oxford University Press, Oxford

Kornberg, A. and Clarke, H. D. (1992) *Citizens and Community: Political Support in a Representative Democracy*, Cambridge University Press, Cambridge

Kumlin, S. (2002) *The Personal and the Political: How Personal Welfare State Experiences Affect Political Trust and Ideology*, Göteborg Studies in Politics, Göteborg

Lidskog, R. and Sundqvist, G. (2002) 'The role of science in environmental regimes: The case of LRTAP', *European Journal of International Relations*, vol 8, pp77–101

Lipset, S. M. and Schneider, W. (1987) *The Confidence Gap: Business, Labour and Government in the Public Mind*, Free Press, New York

Listhaug, O. (1995) 'The dynamics of trust in politicians', in Klingeman, H. D and Fuchs, D. (ed) *Citizens and the State*, Oxford University Press, Oxford

March, J. G. and Olsen, J. P. (1989) *Rediscovering Institutions: The Organizational Basis of Politics*, The Free Press, New York

Merton, R. (1973) 'The normative structure of science', in Merton, R. (ed) *The Sociology of Science: Theoretical and Empirical Investigations*, Chicago University Press, Chicago

Meyer, J. W. (2000) 'Globalization: Sources and effects on national states and societies', *International Sociology*, vol 15, pp233–248

Meyer, J. W. and Jepperson, R. L. (2000) 'The "actors" of modern society: The cultural construction of social agency', *Sociological Theory*, vol 18, pp100–120

Meyer, J. W., Boli J., Thomas, G. and Ramirez, F. (1997) 'World society and the nation-state', *American Journal of Sociology*, vol 103, pp144–181

Petty, R. E. and Cacioppo, J. T. (1986) *Communication and Persuasion: Central and Peripheral Routes to Attitude Change*, Springer-Verlag, New York

Price, D. (1965) *The Scientific Estate*, Harvard University Press, Cambridge, MA

Putnam, R. (1993) *Making Democracy Work*, Princeton University Press, Princeton, NJ

Renn, O. (1992) 'Concepts of risk: A classification', in Krimsky, S. and Golding, D. (eds) *Social Theories of Risk*, Praeger Publishers, Westport

Scholz, J. T. (1998) 'Trust, taxes and compliance', in Braithwaite, V. and Levi, M. (eds) *Trust and Governance*, Russel Sage Foundation, New York

Skodvin, T. (2000) 'The intergovernmental panel on climate change', in Andresen, S., Skodvin, T., Underdal, A. and Wetterstad, J. (eds) *Science and Politics in International Environmental Regimes: Between Integrity and Involvement*, Manchester University Press, Manchester and New York

Weinberg, A. (1972) 'Science and trans-science', *Minerva*, vol 10, pp209–222

Wynne, B. (1980) 'Technology, risk and participation: The social treatment of uncertainty', in Conrad, J. (ed) *Society, Technology and Risk*, Academic Press, London

Wynne, B. (1992) 'Misunderstood misunderstanding: Social identities and public uptake of science', *The Public Understanding of Science*, vol 1, pp281–302

4

Is There a Trade-Off Between Cost-Effective Climate Policies and Political Legitimacy?

Henrik Hammar and Sverker C. Jagers

INTRODUCTION

As indicated in the introductory chapter, the climate change issue can be viewed as a social dilemma. The atmosphere's absorptive capacity makes it a good that provides utterly important services for society and the global biological system by absorbing greenhouse gases, thus regulating the temperature and, over time, also the climate. The atmosphere's accessibility and the non-excludability of its services also make it a public good. Simultaneously, the atmosphere's absorptive capacity is limited. If greenhouse gases are emitted beyond those limitations, the temperature will eventually increase, leading to climatic changes evidently affecting society in a number of negative ways.

Thus, we are facing a social dilemma. In the shorter term, each emitter privately gains the most by unbounded emitting. Practically regardless of the individual emitter's contribution to the emissions, his or her contribution will not make much of a difference to the process of global climate change. The rationale for the emitter of carbon dioxide (CO_2) emissions is thus to continue emitting, while hoping that everybody else is decreasing their emissions. Short-term gains simply seem more attractive than sacrifices for longer-term and often uncertain benefits. Furthermore, many people appear to be willing to take the climate-related risks imposed by their unrestricted behaviour.

Since people are usually reluctant to act freely for the benefit of the collective,[1] one effective way of lessening this social dilemma is through active intervention to change the behaviour that causes it. Such intervention can be executed in many different ways and be more or less painful to those affected. In this chapter we focus on possibilities for steering people's choice and current use of private transportation, which is a major contributor to global CO_2 emissions.

One often-proposed instrument to affect behaviour is information. However, research indicates that information about the negative effects from car driving on the environment is not in itself very effective (Krantz-Lindgren, 2001; von Borgstede, 2002). At the same time, information is the mildest form of governmental intervention and, thus, is relatively appreciated among the public, as well as among politicians. In the case of individual car use, it is reasonable to expect that the perceived advantages of continued driving often outweigh the impact of information about the disadvantages of further emissions and future climate change. However, information can be a prerequisite for other more costly and painful measures to work in an adequate way. Knowledge about the negative environmental effects of their driving may make citizens more susceptible to, and accepting of, other and more forceful measures.

Further investments in public transportation provide another (expensive) way of inducing a less car-dependent behaviour in citizens. Although expensive, it has the advantage of not restricting the individual's freedom of choice. If someone lives in a neighbourhood or a region with inadequate public transportation, more frequent buses or trains would provide a choice and an incentive for people to refrain from using the car as the single most important means of conveyance.

The most powerful intervention tool to decrease car driving is legislation and prohibition. However, upholding such a system is costly, since legislation veering on prohibition or rationing carries with it high administrative costs and may lead to a loss of legitimacy as people feel their freedom of choice (unduly) circumscribed.

Seen from the perspective of both liberty and economic costs, market-based policy instruments such as subsidies and taxes are among the most interesting options. Subsidizing measures can be relatively expensive; but they do not deprive much individual freedom. However, what about 'negative' economic interventions, such as taxes on carbon dioxide? Taxes are typically cost-effective measures for society. Furthermore, they still offer a certain freedom of choice, although this is dependent upon individuals' level of income and the possibility of avoiding tax (for example, by filling your car's petrol tank in a neighbouring country).

In this chapter, we focus on the Swedish carbon dioxide tax, which was first introduced in 1991. It is today the main climate policy tool for regulating CO_2 emissions from private transport in Sweden. Our main reason for studying this tax is that it has a number of environmental, as well as societal, advantages compared to most other climate policy tools. First, the CO_2 tax forces the individual *polluter* to pay for the environmental costs caused by his or her consumption, while most other available instruments force all *taxpayers* – regardless of their contribution to the overall emissions – to contribute. Second, compared with the rather self-sufficient system of the CO_2 tax, practically all other policy options presuppose comprehensive investments (in, for example, labour resources and control mechanisms). In other words, making use of the CO_2 tax is a simple and cost-effective way to affect people's transportation behaviour. Despite these seemingly positive characteristics, the Swedish CO_2 tax is nevertheless an utterly unpopular policy instrument (Hammar and Jagers, 2003). This warrants us to expand on *why the well-functioning CO_2 tax is so unpopular*.

In order for climate policies to be both economically efficient and politically feasible, one should focus on policies that contribute to high cost-effectiveness and public support. In three different mail surveys, we addressed this matter by asking the Swedish general public about their opinions on different measures to reach climate policy goals.[2] A general interpretation from the three surveys is a response pattern of 'passing on costs' to other taxpayers – that is, an indication of the importance of self-interest.[3] Assume that you are consuming a lot of fossil fuel for private transport. From an economic incidence perspective, it is then better that emissions are reduced by other means than a CO_2 tax since this implies a lower private tax burden (no increase in the CO_2 tax and just a fraction of other tax increases since they are typically shared by all taxpayers). Another aspect of the attitude surveys is that they are typically one dimensional in the sense that there is no incentive for the respondent to make trade-offs between, for instance, a decreased tax on green fuels and an increase of some other tax to finance the green fuel tax subsidy.[4]

Figure 4.1 illustrates how we perceive the relationships among policy, attitude and mediating (e.g. justice and trust) variables. The arrows with solid lines represent the aspects of the model that we primarily deal with in this chapter. We also acknowledge that there can be a direct effect of policy on policy attitude (A). When viewed from a societal perspective, we can also expect a feedback (B) effect of attitude on policy (i.e. affecting the future shape of the policy). Furthermore, as can be seen in the upper part of the figure (C and D), we assume that both policy (e.g. by providing incentives for behavioural change) and attitude towards policy (e.g. policy triggers certain values that are, in different ways, in line with policy characteristics) can affect individual behaviour, in this case encouraging them to change or reduce their driving.

In the study, we proceed from a hypothetical suggestion to increase the CO_2 tax and then concentrate on how the potential explanatory and mediating factors (self-interest, justice and trust) strengthen or weaken the respondents' attitudes to the suggested policy change. Two analyses are performed. The first focuses on how preferences for fair emission reductions affect the support for increases in the CO_2 tax, and whether such preferences can alleviate the weight of self-interest. The second discusses whether and how trust in politicians can explain the support for a tax increase. The two analyses use data from two different surveys and are partly based on previous work (Hammar and Jagers, 2006, forthcoming). Instead of singling out and studying the effect of self-interest in isolation, we control for it when analysing the significance of fairness and trust: do these two factors reduce the expected impact of self-interest? Our results indicate that both preferences for fairness and trust in politicians tend to benefit 'the common good' since both increase the likelihood that

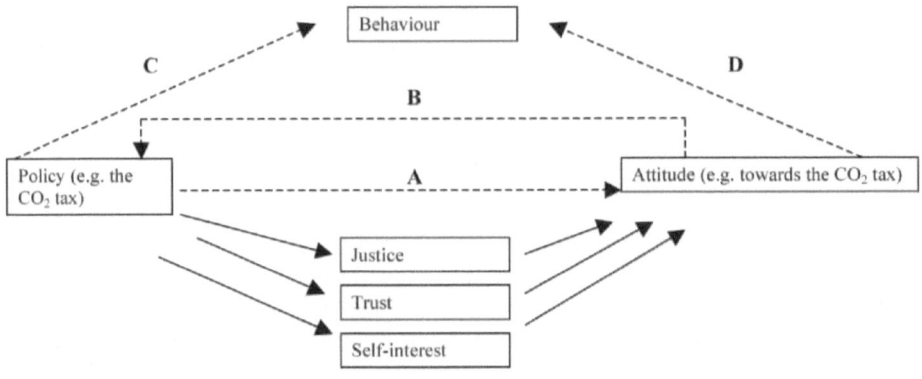

Source: chapter authors

Figure 4.1 *Schematic description of the relationship among attitudes, behaviour and policy*

people will support suggested increases in the CO_2 tax. Consequently, they weaken the effect of self-interest.

Self-Interest Versus the Common Good: The Case of the CO_2 Tax

Before presenting the analyses, it is worthwhile summarizing our measurements of fairness, trust and self-interest. With *fairness* we refer to preferences concerning the incidence of emission reductions. We create an index of this preference from individual responses to three different questions – theoretically corresponding to different fairness principles – on fair emission reductions. *Trust* here refers to trust in politicians since they have the power to introduce, as well as to change, the CO_2 tax. We measure this form of trust by the question: 'Generally speaking, how great is your trust in Swedish politicians?' *Self-interest* refers to three things:

1 car dependence (i.e. frequency of car use or access to a car in a household);
2 perceptions of climate change as a threat; and
3 place of living, which we hypothesize to be relevant for car dependence.

The importance of fairness preferences

One objection against a CO_2 tax is that it implies unjustly heavy burdens for the CO_2 emitters. This is a reasonable objection because the CO_2 tax increases the gasoline price. In turn, this primarily affects low-income earners and individuals living in remote areas since these groups spend a comparatively large share of the household budget on private transportation (Kriström et al, 2003). Here we analyse whether preferences for fairness matter regarding support for increases in the Swedish CO_2 tax and whether these preferences have a counteracting effect on the importance of self-interest.

The importance of justice and people's perceptions of fair distribution has occupied empirically oriented scholars from a range of disciplines (Sen, 1988; Mansbridge, 1990; Young, 1995; Tyler et al, 1997; Douglas, 1998; Roemer, 1998; Kolm, 2002; Jagers, 2006). A variety of matters are of concern in these studies. Some focus on people's views of justice and fairness, and others on whether these views can serve as explanations, such as whether conceptions of justice can explain (positive and negative) political attitudes (Tyler et al, 1997) and whether views of justice matter for people's behaviour in interaction with others (e.g. Eek, 1999). Below we analyse the importance of preferences for fair emission reductions with a fairness index based on three distributional principles: equality, need and equity.

The data used for this analysis originates from a questionnaire sent out to a random sample of 2000 individuals, aged 18 to 85, drawn from the Swedish population in the National Register. The net response rate was 58 per cent, or 1133 respondents. Due to item non-response on various questions, 932 respondents were used in the estimations.[5]

Table 4.1 shows the distribution of responses to the questions of fair reductions, tied to the greenhouse effect. The most obvious pattern is that respondents apprehend all principles to be 'somewhat fair' or 'very fair', ranging from 73 per cent for the *equity principle* (people who pollute the most should decrease their emissions more than others) down to 46 per cent for the *equality principle* (everybody should decrease their emissions the same – for example, by 10 per cent). However, the respondents also make a distinction between the fairness principles. The principle that really stands out is equity: only 8 per cent

Table 4.1 *Distribution of responses regarding fair reductions of CO_2 emissions from private car transport*

Principle	I believe that …	Very fair	Some- what fair	Neither fair nor unfair	Some- what unfair	Very unfair	No opinion	No res- ponse	Opinion balance
Need	People who need their cars the least (e.g. who have access to public transportation) should decrease their emissions more than others	24%	28%	19%	13%	10%	4%	2%	29%
Equality	Everybody should decrease their emissions the same (e.g. by 10%)	20%	26%	21%	14%	12%	4%	2%	20%
Equity	People who pollute the most should decrease their emissions more than others	43%	30%	12%	5%	3%	4%	2%	65%

Note: total population sample size (n) = 1106. The opinion balance measure shows the share of respondents who find the respective fairness principles very fair or somewhat fair, *minus* those who think they are somewhat unfair or very unfair.

think it is 'somewhat unfair' or 'very unfair', and, furthermore, only 16 per cent have 'no opinion' or believe it is 'neither fair nor unfair'. Hence, the equity principle appears to be considered by people as the most fair when distributing emission cuts.

This response pattern also indicates that there are other aspects besides cost-effectiveness that appear to be important determinants for individuals' attitudes. However, the most popular fairness principle is in line with the use of a CO_2 tax since policies using a polluter pays principle (PPP) tend to imply that those who emit the most also have an incentive to reduce their emissions the most. However, this must not always be the case because people have different incomes and vary in their dependence upon fossil fuels. Indeed, if the question were asked in a global justice perspective, then the equity principle would most likely not be in line with a general application of PPP because of the uneven income distribution in different countries. In a Swedish context, however, those preferring that the largest emitters should decrease emissions the most can still be expected to be positive towards an application of PPP in the form of a CO_2 tax.

To make different fairness principles operational and to frame them as fair reductions of CO_2 emissions from private car transportation poses a great challenge. Therefore, we are the first to acknowledge that our implementation of fairness principles is coupled with measurement problems. This is also an important rationale for creating an index of the responses.[6] It is also important to bear in mind that, to some extent, the questions of fairness in allocating reductions measure attitudes towards reductions in CO_2 emissions. Hence, the focus on the response patterns in Table 4.1 should primarily lie in the *difference* between those who consider the fairness principles fair and unfair – which is an additional reason for creating an index of these three questions. From the responses we create the index variable, which we categorize into three groups. The highest value of what we label 'fairness intensity' is 15, corresponding to the case when a respondent says that all three principles for emission reductions are 'very fair'. The other end point is 3: all three principles are 'very unfair'. In the estimations below, we use the following three fairness categories:

1   'fairness intensity' low = 1 for fairness values below 10; 0 otherwise – corresponding to 33 per cent of the observations (comparison group in estimations);
2   'fairness intensity' middle = 1 for fairness values 10 and 11 (15 maximum); 0 otherwise – corresponding to 32 per cent of the observations;
3   'fairness intensity' high = 1 for fairness values above 11 (15 maximum); 0 otherwise – corresponding to 35 per cent of the observations.

Table 4.2 presents econometric results from this approach. Looking at model 1, it can first be seen that younger age groups are less supportive than older age groups (ages 50 to 85). This is also true for people living in the countryside and for frequent car users. The effect of car use on the support for increases in the CO_2 tax is relatively strong, and is only stronger for Green party sympathizers, even if this 'stronger' effect is positive in terms of support. We also see that being female, having studied at university and being single implies slightly less support for increases in the CO_2 tax.

Since the burdens to be distributed are CO_2 reductions from private car transportation, thus potentially evoking self-interest, it is important to see whether the group of daily car users differs from those using their cars less frequently. One obvious reason and explanation for such a difference is that the former would carry a larger tax burden than the latter. From a self-interest point of view, it appears that someone who uses private car transport on a regular basis would be more likely to vote 'no'. However, there is no perfect relation. Even among the regular car users, there are those who would vote in favour of increasing the CO_2 tax. Yet, it should be stressed that the answers here do not allow for interpretations in monetary terms since 'daily use' could mean a few, as well as many, kilometres per day.

Second, when comparing the results in models 1 and 2, we see the significance of fairness in emission reductions. Respondents with high fairness intensity tend to be much more supportive towards increases of the CO_2 tax than those with middle or low intensity.

Third, when splitting the sample (models 3a and 3b) into infrequent and frequent car users (i.e. not 'forcing' the same probability distribution on the two sub-groups), we see that the explanations for the support towards increases of the CO_2 tax can be distinguished further. For instance, among infrequent car users, the 'Green party' dimension becomes even stronger, while it (statistically speaking) disappears for frequent car users. This is presumably partly an effect of the low number of Green party sympathizers being frequent car users.

Table 4.2 *Parameter estimates from logit model: Dependent variable 'vote in favour of increased CO_2 tax' compared to 'vote against increased CO_2 tax'*

	Model 1 without fairness preference	Model 2 with fairness preference	Model 3a: infrequent car users	Model 3b: frequent car users
'Fairness intensity' middle	—	0.285	0.062	0.470
'Fairness intensity' high	—	0.884***	0.693**	1.101***
Ages 18 to 29	-0.708***	-0.607**	-0.707**	-0.460
Ages 30 to 49	-0.507***	-0.417**	-0.420	-0.363
Female	0.326*	0.252	0.439*	0.095
Studies at university	0.295*	0.427**	0.304	0.620**
Single	0.339*	0.353*	0.422	0.154
Green party sympathizer	1.387***	1.292***	1.888***	0.455
Frequent car user	-0.848***	-0.800***	—	—
Lives in the countryside	-0.607***	-0.644**	-1.138*	-0.397
Lives in a big city	0.186	0.247	0.132	0.559
Constant	-0.889***	-1.421***	-1.326***	-2.398***
Log likelihood	-458.3	-448.9	-221.9	-222.9
Number of observations	932	932	379	553

Notes: * $p < 0.10$; ** $p < 0.05$; *** $p < 0.01$.

Furthermore, fairness seems to be of greater importance, judging from the size of the coefficients for frequent car users. Nevertheless, the effect of fairness is significant for both sub-groups.

When political interventions are contemplated in order to avoid or mitigate social dilemmas, attention must be paid to the distributional effects of the policy instruments. Obviously, perception of fairness is one factor that influences the public's attitudes towards potential increases of the present Swedish CO_2 tax. In case further increases are planned – and, in particular, if gathering further public support for such an increase is desirable – a balancing of the CO_2 tax increase against, for instance, decreased income taxes may be one way of gaining support among lower income groups. This can be made in combination with information campaigns directed towards the high-emitting group with the primary purpose of introducing the logic behind the polluter pays principle. Simply put, there is an (environmental) reason for why it becomes more and more expensive to drive a car.[7]

The importance of trustworthy politicians

After discussing the importance of preferences for fair emission reductions, we now turn to the other main factor that may affect people's attitudes towards an increased CO_2 tax: *trust in politicians*.[8] We offer two reasons why political trust may explain that some people tend to accept or support an increased CO_2 tax while others do not.

The first reason is quite simple: why would people support an increased tax burden imposed on them by actors or institutions whom they do not trust? The second, related, reason is the very nature of the CO_2 tax. It is not just a source of revenue; it is also (primarily) a steering instrument. The public must thus be confident that the tax will have the intended effects and be relatively certain that the government will use the revenues in an appropriate way – that is, for a proper cause.

As we can see from Table 4.3, very few respondents display a 'very strong trust' in politicians. Furthermore, the responses are asymmetrical in the sense that the distribution is tilted towards weak political trust. In the regressions (see Table 4.4), we merge those with very or rather strong trust and those with rather or very weak trust, respectively.[9]

Table 4.3 *Distribution of responses of trust in politicians (percentage)*

Generally speaking, how strong is your trust in Swedish politicians?	
Very strong trust	1
Rather strong trust	35
Rather weak trust	53
Very weak trust	10

Note: total population sample size (n) = 1345.

Table 4.4 *Explanations of a positive attitude towards increases in the CO_2 tax on petrol: Changes in probabilities for a discrete change of dummy variable from 0 to 1, evaluated at sample mean, based on logit estimation*

	Model 1: full sample	Model 2: full sample, trust excluded	Model 3a: high political trust	Model 3b: low political trust
Trust in politicians	0.067***	—	—	—
Male	−0.036	−0.033	−0.018	−0.049*
Living in city	0.038	0.040	0.075*	0.022
Left wing	0.052*	0.062*	0.044	0.048
Right wing	−0.010	−0.010	0.061	0.010
Green party sympathizer	0.146***	0.151***	0.220***	0.097*
High school educated	0.029	0.032	0.086	−0.005
Post-high school educated	0.086***	0.092***	0.168***	0.038
Ages 18 to 30	−0.043	−0.044	−0.095	−0.016
Ages 31 to 60	−0.073**	−0.076**	−0.068	−0.076**
Access to car	−0.102***	−0.105***	−0.040	−0.141***
Consider climate change as a threat	0.062**	0.059**	0.090**	0.046*
Taxes are a very effective tool to change individual behaviour	0.180***	0.187***	0.163***	0.196***
Log likelihood	−578.8	−582.9	−245.1	−326.8
Number of observations	1270	1270	462	808

Notes: * $p < 0.10$; ** $p < 0.05$; *** $p < 0.01$.
Source: adapted from Hammar and Jagers (2006)

First, it is clear that there are factors other than trust that are more important in terms of the size of the effect. Measured by changes in probabilities of having a positive attitude, the effects (when looking at model 1 of whether one believes that taxes are an effective way of changing people's behaviours, the Green party dimension, access to a car, and people with high education) are stronger than trust in politicians in terms of size. Thus, *lack of* trust in politicians should be added to the list of other factors contributing to a fuller understanding of the massive opposition against (or the moderate support for) increases in the CO_2 tax.

Second, when comparing the Green party sympathizers with all other party sympathizers, the green dimension of the CO_2 tax issue comes out quite clearly. It also turns out that respondents who consider themselves to be 'left wing' are more positive compared to 'right-wing' respondents and those in the middle. When comparing respondents with and without access to a car, those having access are less likely to support a CO_2 tax increase. This result is expected according to previous research on self-interest, since those who have cars would suffer considerably more from increased costs of car ownership and use.[10] The same is true for those who view climate change as a great risk. For selfish (and presumably also collective) reasons, they are more likely to support tax increases.[11]

Third, beliefs that a policy measure would have the intended effects may increase the probability that people will be more willing to accept its implementation. In the case of a CO_2 tax, two different but kindred beliefs are plausible: changes in human behaviour and/or societal changes with more immediate positive environmental effects. The question that was asked in the survey captures the first conception of effect and increases the probability of supporting a tax increase. This could imply that one way for the Swedish government to implement an increased CO_2 tax might be a nationwide campaign aimed at persuading the Swedes that such a tax actually 'does the job'.

Fourth, with regard to the demographic factors, the 31 to 60 age group is more negative than other age groups, and support for CO_2 tax increases is not appreciably affected by gender or by where respondents live (this indicates some conflicting evidence of support for increased CO_2 taxes when it comes to demographic factors; compare Table 4.2). As for education, it is more likely that higher educated people (students or former students at universities/colleges) have positive attitudes towards a CO_2 tax increase than less educated people.

Finally, we gain additional understanding by looking at the support for an increased CO_2 tax on petrol by splitting up the respondents into high-trusting and low-trusting groups. High-trusting 'green' individuals are more likely to support CO_2 tax increases than low-trusting 'greens'. Thus, we here see a sign of a 'trust' dimension in green political issues, implying that support for climate policies is not a mere green issue. Furthermore, the group which according to the self-interest literature ought to be most fervently against an increased CO_2 tax is the one with access to a car. From the perspective of trust as a potential counteragent to self-interest, it is thus promising to find that this opposition is only valid for low-trusting people with access to a car. Motorists who trust their politicians are not more likely to resist CO_2 tax increases than high-trusting persons with no access to a car. However, it might be problematic to extrapolate this finding. For instance, if the CO_2 tax were to be increased quite substantially, trust may become less important relative to the self-interest of a 'reasonable' tax burden.

CONCLUSIONS

We began the chapter by arguing that the case of climate change is a typical example of a social dilemma situation. We also contended that the only really effective way to avoid or mitigate (but not necessarily *overcome*) this situation is through active regulation of human behaviour. By posing disincentives to people making large-scale emissions, the longer-term collective costs can be kept as low as possible. However, although the CO_2 tax has been in operation since 1991, it has not resolved the dilemma. The conflict between self-interest and public interest is thus still present. While people are supportive of measures aimed at improving our environment and climate, this support is reduced as soon as people have to make sizeable efforts and personal sacrifices.

Let us now connect this line of reasoning with the core question of the chapter: why is something as well functioning as the Swedish CO_2 tax so unpopular? First, even if this stylized fact still gains support in our study (since an increased CO_2 tax typically presupposes noticeable behavioural changes that will cost money, comfort, availability or time at the individual level), we find that this 'standard' answer has to be modified when preferences for fair emission reductions and trust in politicians are considered. Our results indicate that although the CO_2 tax is unpopular, both preferences for fairness and trust in politicians tend to benefit 'the common good' since both these factors seem to increase the likelihood that people will support tax increases.

Bearing in mind that attitude surveys have limited value in political debate, an informed discussion on how the potential support and non-support vary among different groups is valuable. A negative attitude can potentially be alleviated if policies are adapted in such a way that certain values become less pronounced (see Chapter 5 in this volume). Thus, we believe that our results can serve as a tool to perform a more informed discussion about present and future climate policy.

Let us now consider the unpopularity of the CO_2 tax from a slightly different angle. As we have contended throughout the chapter, a CO_2 tax is both environmentally effective at current tax levels (Ds 2001:71) and cost-effective for combating climate change (Söderholm and Hammar, 2005). Judging from public opinion, such cost-effective climate policies are still unpopular. The immediate distributional effects on the individual household economy are probably one reason for this unpopularity (see, for example, Kriström et al, 2003). Other much more popular policy instruments among the public (e.g. expanded public transport) are, however, both less environmentally effective and less cost-effective, and they also require unpopular financing via other taxes (see Hammar et al, 2005).

Hence, another challenging finding in our study is the apparent juxtaposition of cost-effectiveness and justice. This is rather interesting. When respondents are asked only to give their opinions about different climate change policy measures, they are much more sympathetic towards alternative instruments such as non-fossil fuel subsidies and expanded public transportation. However, when responding to these kinds of questions, it is reasonable to ask whether people really contemplate their individual costs in terms of the necessary increase of other taxes (to finance those alternative measures). How much would an expanded public transportation system in Sweden cost in terms of further taxation? How much would comprehensive fuel subsidies cost? When comparing responses from different surveys, we have reason to believe that respondents tend to give quite 'one-dimensional' responses – that is, they do not fully take in the total costs or the costs for alternative measures.

Interestingly enough, when comparing the popularity of increasing the CO_2 tax with the popularity of increasing *other* taxes – that is, fiscal taxes and other taxes that would be increased in order to finance, for example, fuel subsidies – it turns out that the CO_2 tax is a relatively popular tax. In fact, of 11 taxes, respondents rank the CO_2 tax as the fourth most acceptable to increase, coming right after the wealth, payroll and corporate taxes (Hammar et al, 2005).

It seems reasonable to expect that the perception of a fair tax system acknowledges the difference in tax rates and tax burden, and whether the tax, as such, is considered to be a 'fair' tax. For instance, the Swedish real estate tax is a very unpopular one, possibly due to the fact that the tax costs for some groups of high concern (e.g. elderly people) are considered too substantial. This is in spite of the fact that this is an excellent tax if one only has optimal taxation in mind. In contrast, it appears as if the CO_2 tax should be considered fair. It is an equitable tax in line with the polluter pays principle, and should primarily be seen from a distributive efficiency perspective – that is, a tax with the goal of reaching an efficient allocation of scarce resources. A reasonable interpretation from a fairness perspective is that those who emit the most should also decrease their emissions the most. Nevertheless, this is not a view held by all groups as it is less common among those who use their car on a daily basis. Given the present construction of the tax, this group happens to carry a higher tax burden.

Without claiming to have any final answer, we believe that it is worth discussing the actual relevance of the unfairness argument against the Swedish CO_2 tax and, thus, how to weigh the fairness and efficiency aspects of that tax. Perhaps there should not be any conflicting interests? First, how unfair is it to have actual polluters pay for their contribution to the climate-related social dilemma situation, compared with having all tax-paying Swedes pay the bill for expanded public transportation? Second, since practically all Swedes, regardless of how much they drive and earn per year, both emit and earn considerably more per capita than a majority of the world's population, the Swedish CO_2 tax is arguably justified in a global equity perspective.

NOTES

1   According to Olson (1965), the problem of free-riding, which is a risk when it comes to achieving an efficient allocation of society's resources, tends to increase with group size. The explanation is simply that the smaller the individual's ability to affect an outcome that is dependent upon the actions of others, the more probable it is that the individual will defect from cooperation. This probability tends to increase as the relevant population increases in number. This phenomenon is apparent in many aspects of environmental politics such as climate policy, and can be exemplified by statements such as 'Why should Sweden be so good when the US has not even ratified the Kyoto Protocol?' and 'Why should I use public transport while "all" others use their cars?' Hence, when our own climate-friendly action is perceived as more or less meaningless due to the behaviour of others, then the individual's defection (e.g. in the form of driving an extra kilometre) from the collectively best option can be interpreted as a sign that self-interest is considered the best option. This is, indeed, a challenge for policy-making, but also a rationale for governmental coordination.

2   The first and third surveys were sent out by the Society-Opinion-Media (SOM) Institute at Göteborg University in the autumn of 2002 and 2004 to a random sample of 3000 individuals from the Swedish population, aged 18 to 85, collected from the National Register. The second survey was administered by ourselves, where a random sample was drawn from the Swedish population aged 18 to 85 in the National Register. This questionnaire was sent out to 2000 individuals. The net response rate was 58 per cent, or 1133 respondents.

3   Self-interest as a predictor for support – or non-support – of policies is not a new idea. Comprehensive social science literature points out individuals' self-interests as a factor that largely explains attitudes towards public authorities, parties and singular political propositions (Lind and Tyler, 1988; Sears and Funk, 1990; Tyler 1990a, 1990b; Dunleavy, 1991). The more a policy satisfies, or promises to satisfy, individuals' economic or political preferences, the more positive attitudes people tend to have towards it.

4   It is important to stress that we are measuring attitudes in this chapter, and we analyse how certain attitudes, especially regarding the CO_2 tax, can be explained by a range of background factors. However, it requires courageous assumptions for someone to directly interpret these attitudes as individuals' real preferences for a particular policy since attitude surveys typically ask questions in one dimension, which makes it impossible to draw conclusions from, for instance, a policy package where a CO_2 tax increase can be coupled with a decrease in the labour tax.

5   In the group that has no opinion about whether the CO_2 tax should be increased (10 per cent, or 115 responses), females tend to be heavily over-represented (70 per cent) and people with university education tend to be somewhat under-represented (30 per cent, compared to 36 per cent in the whole sample). The latter fact is consistent with the thought that more-informed citizens (measured by education) tend to be more likely to have an opinion.

6   However, see Hammar and Jagers (forthcoming) for an analysis where the principles are also treated as independent explanatory variables.

7   In the data, we also have some, though limited (due to a large non-item response on the income question), information on income, which is why we choose not to include this in the regressions. However, in a reduced sample (not presented in Table 4.2), we do not find any statistically significant effect from whether a household is low or high income.

8   Here we make use of data from a nationwide survey annually sent out by the SOM Institute at Göteborg University. Different interpretations/implementations of trust are becoming increasingly important factors in explaining phenomena such as why people tend to pay tax, although rational choice assumptions typically urge us not to (Scholz and Pinney, 1995; Scholz and Lubell, 1998; Feld and Frey, 2002). Over the last 15 years, the idea of social capital has earned more and more interest within the social sciences. Although contested (Portes, 1998; Fine, 2001; Bankston and Zhou, 2002; Robison et al, 2002), social capital usually refers to two different aspects: degree of social networking and degree of trust in people and officials such as politicians (see, for example, Putnam et al, 1993; compare Wuthnow, 1998; Stolle, 2001). 'Social networkers' are then socially skilful people with experience from, for example, associations and club activities. These people, it is argued, are more inclined to accommodate, accept and support the formal rules of society. In principally the same way, interpersonal or generalized trust – to be distinguished from particular trust (Yamagishi and Yamagishi, 1994) – is argued to be an important reason for why people follow and legitimize the institutions of society (Uslaner, 2002). Thus, it is true that groups of outlaw bikers are characterized by both a large degree of social networking/ association activities, as well as by interpersonal trust within the group – that is, particular trust. However, they lack trust in strangers, and while they can be said to do a lot of things, few would accuse them of legitimizing the formal rules and other foundational institutions of society. Furthermore, according to Uslaner (2002), public willingness to conform to policies aimed at dealing with collective action problems is more likely to be found in countries with a relatively high degree of generalized trust. Thus, from a macro-perspective, if most people in a country both participate in social networking and have a large degree of generalized and political trust, then the theory of social capital predicts that the democratic system and the fundamental institutions in that country will work rather well (Dekker and Uslaner, 2001).

9   It is worthwhile pointing at a potential covariance between trust in politicians (attitude object) and what the politicians do (in this case, a potential increase of the CO_2 tax). In practice, survey responses to these two theoretically different factors could measure more or less the same thing, unless respondents differentiate between their view of politicians and what they potentially might implement.

10   A more precise measure of car use, which would have been even more related to self-interest, would have been to use a measure that accounts for kilometres driven, fuel efficiency and to control for income and whether it is a company car or not. When trying to account for this, we find that these measures are too imprecise or have been too difficult to answer (non-item responses). This is why these potentially important measures have been omitted in this analysis.

11   The reason why this effect might seem small may depend upon the small variation in the responses. According to the responses, a lot of things appear to be threatening – not only climate change. Thus, one interpretation is that the different threat questions are actually measuring the same phenomena (compare von Borgstede and Lundqvist, 2002).

REFERENCES

Bankston, C. L. and Zhou, M. (2002) 'Social capital as process: The meanings and problems of a theoretical metaphor', *Sociological Inquiry*, vol 72, pp285–317

Dekker, P. and Uslaner, E. (eds) (2001) *Social Capital and Participation in Everyday Life*, Routledge, London

Douglas, M., Gasper, D., Ney, S. and Thompson, M. (1998) 'Human needs and wants', in Rayner, S. and Malone, E. L. (eds) *Human Choice and Climate Change, Vol 1: The Social Framework*, Battelle Press, Columbus

Ds 2001:71 (2001) *Sveriges tredje nationalrapport om klimatförändringar – I enlighet med Förenta Nationernas ramkonvention om klimatförändringar*, Miljö- och samhällsbyggnadsdepartementet, Stockholm

Dunleavy, P. (1991) *Democracy, Bureaucracy and Public Choice: Economic Explanations in Political Science*, Harvester Wheatsheaf, Hemel Hempstead

Eek, D. (1999) *Distributive Justice and Cooperation in Real-Life Social Dilemmas*, Department of Psychology, Göteborg University, Sweden

Feld, L. P. and Frey B. S. (2002) 'Trust Breeds Trust: How Taxpayers are Treated'. *Economics of Governance*, vol 3, pp87–99

Fine, B. (2001) *Social Capital versus Social Theory: Political Economy and the Social Science at the Turn of the Millennium*, Routledge, London and New York

Hammar, H. and Jagers, S. C. (2003) 'Om svenska folkets acceptans av en skärpt klimatpolitik', in Holmberg, S. and Weibull, L. (eds) *Fåfängans marknad*, SOM-rapport nr 33, SOM Institute, Göteborg University, Sweden

Hammar, H. and Jagers, S. C. (2006) 'Can trust in politicians explain individuals' acceptance of public policies? The case of climate change policy', *Climate Policy*, vol 5, pp611–623

Hammar, H. and Jagers, S. C. (forthcoming) 'What is a fair CO_2 tax increase? On fair emission reductions in the transport sector', forthcoming in *Ecological Economics*

Hammar, H., Jagers, S. C. and Nordblom, K. (2005) 'Skatter och skattefusk', in Holmberg, S. and Weibull, L. (eds) *Lyckan kommer, lyckan går*, SOM-rapport no 36, SOM Institute, Göteborg University, Sweden

Jagers, S. C. (2006) *Prospects for Green Democracy*, University Press of America, Lanham, MD

Kolm, S.-C. (2002) *Justice and Equity*, MIT Press, Cambridge, MA

Krantz-Lindgren, P. (2001) *Att färdas som man lär*, PhD thesis, Göteborg Studies in Political Science, Göteborg University, Sweden

Kriström, B., Wibe, S., Brännlund, R. and Nordström, J. (2003) 'Fördelningseffekter av miljöpolitik', bilaga 1 in *Långtidsutredningen*, Ministry of Finance, Stockholm (in Swedish)

Lind, A. E. and Tyler, T. R. (1988) *The Social Psychology of Procedural Justice*, Plenum Press, New York

Mansbridge, J. J. (1990) *Beyond Self-Interest*, Chicago University Press, Chicago

Olson, M. (1965) *The Logic of Collective Action: Public Goods and the Theory of Groups*, MIT Press, Cambridge, MA

Portes, A. (1998) 'Social capital: Its origins and applications in modern sociology', *Annual Review of Sociology*, vol 24, pp1–24

Putnam, R. D. with Leonardi, R. and Nanetti, R.Y. (1993) *Making Democracy Work: Civic Traditions in Modern Italy*, Princeton University Press, Princeton, NJ

Robison, L. J., Schmid, A. A. and Siles, M. E. (2002) 'Is social capital really capital?', *Review of Social Economy*, vol 60, no 1, pp1–21

Roemer, J. E. (1998) *Theories of Distributive Justice*, Harvard University Press, Cambridge, MA

Scholz, J. T. and Lubell, M. (1998) 'Trust and taxpaying: Testing the heuristic approach to collective action', *American Journal of Political Science*, vol 42, pp398–417

Scholz, J. T. and Pinney, N. (1995) 'Duty, fear and tax compliance: The heuristic basis of citizenship behavior', *American Journal of Political Science*, vol 39, pp490–512

Sears, D. O. and Funk, C. L. (1990) 'Self-interest in Americans' political opinions', in Mansbridge, J. J. (ed) *Beyond Self-Interest*, Chicago University Press, Chicago

Sen, A. (1988) 'Freedom of choice: Concept and content', *European Economic Review*, vol 32, pp269–294

Söderholm, P. and Hammar, H. with contributions from Berg, C. and Spendrup Thynell, T. (2005) *Kostnadseffektiva styrmedel i den svenska klimat- och i energipolitiken? Metodologiska frågeställningar och empiriska tillämpningar*, ER 2005:30, Energimyndigheten (STEM), Eskilstuna, Sweden

Stolle, D. (2001) 'Getting to trust: An analysis of the importance of institutions, families, personal experiences and group membership', in Dekker, P. and Uslaner, E. (eds) *Social Capital and Participation in Everyday Life*, Routledge, London

Tyler, T. R. (1990a) 'Justice, self-interest, and the legitimacy of legal and political authority', in Mansbridge, J. J. (ed) *Beyond Self-Interest*, Chicago University Press, Chicago

Tyler, T. R. (1990b) *Why People Obey the Law: Procedural Justice, Legitimacy, and Compliance*, Yale University Press, New Haven

Tyler, T. R., Boeckmann, R. J., Smith, H. J. and Huo, Y. J. (1997) *Social Justice in a Diverse Society*, Westview Press, Boulder, CO

Uslaner, E. M. (2002) *The Moral Foundations of Trust*, Cambridge University Press, Cambridge

von Borgstede, C. (2002) 'Miljöengagemang och miljöhandlingar i psykologisk belysning', in *Att handla rätt från början – En kunskapsöversikt om hur konsumtions- och produktionsmönster kan bli mer miljövänliga*, Naturvårdsverket Rapport 5226, Stockholm

von Borgstede, C. and Lundqvist, L. J. (2002) 'Nytt klimat för miljöpolitiken?', in Holmberg, S. and Weibull, L. (eds) *Det våras för politiken*, SOM-rapport 30, SOM Institute, Göteborg University, Sweden

Wuthnow, R. (1998) *Loose Connections: Joining Together in America's Fragmented Communities*, Harvard University Press, Cambridge, MA

Yamagishi, T. and Yamagishi, M. (1994) 'Trust and commitment in the United States and Japan', *Motivation and Emotion*, vol 18, pp129–166

Young, P. H. (1995) *Equity in Theory and Practice*, Princeton University Press, Princeton, NJ

5

Assessing Values in Environmental Policy Formation

Andreas Nilsson and Anders Biel

INTRODUCTION

Scientific efforts to understand global environmental changes on land, in the oceans and in the biosphere began among atmospheric chemists, meteorologists, ecologists and other natural scientists. Because of the anthropogenic nature of current global change, it was, however, soon realized that social and behavioural sciences could contribute to an understanding of how to bring about changes in human behaviour necessary to prevent or respond to global change. Most of the early research efforts by psychologists were derived from psychological concepts in relative isolation from the practical issues of policy-making. However, we believe that a successful implementation of policy instruments requires an understanding of the motivations that determine people's attitudes towards such policy instruments. In our view, a more applied approach in which the practical issues are taken into account should be sought. We thus propose that psychology, together with natural and other social sciences, should work closer to the policy process. In this endeavour, the contribution from psychology should not come to a halt at the individual perspective in policy-making, but should also try to relate this perspective to other social scientific levels of analysis.

That most policy instruments require some kind of sacrifice from the receiving part, either economically (e.g. environmental taxes) or in terms of behavioural change (e.g. reduced car use), may evoke negative attitudes towards these instruments. Psychological research has, however, shown that attitudes are not only based on expected personal consequences. A wider range of factors might motivate people to express themselves in the positive or in the negative about an issue (see Eagly and Chaiken, 1993). When attitudes concern environmental issues, moral motives have to be taken into account. Values and norms have, in previous studies, been found to be important constructs in guiding people's moral decisions (Schwartz, 1977; Kerr et al, 1997), as well as their environmental decisions (Karp, 1996; Fransson and Gärling, 1999; Stern et al, 1999; Nordlund and Garvill, 2002). In this chapter, we highlight their importance for shaping people's reactions to policy measures.

In this attempt, values will be of prime interest. After a brief review of the value concept, we present the relationship between values and environmental behaviour on an individual level. Recognizing that individual behaviour also evolves in a social context, we then address the acceptance of policy measures in an organizational context. In the next section we shift focus from the personal and organizational values to the resource being valued. In this section we ask the question: how can personal values be transformed to express a value of a specific resource? We end by discussing some policy implications.

VALUES

Although value research has suffered from some definitional inconsistency (see Rohan, 2000, for a review), earlier and contemporary value theorists seem to agree that human values state what is important in people's lives, that they are conceptions of a desired goal or end state, and that they influence and guide evaluations and behaviour (Kluckhohn, 1951; Rokeach, 1973; Schwartz, 1992; Feather, 1996).

An important contribution to the empirical study of value theory was the explanation and naming of values (Rokeach, 1973). Rokeach asked people to arrange value words in order of their relative importance. This method of measurement, to ask people to rate values in order of importance to you as guiding principles in your life, is now the most commonly used method in value research (Sagiv and Schwartz, 1995). A noticeable limitation with Rokeach's work, however, was that the selection of values was not supported by an underlying theory. The set of values was based on intuition, and there was no theoretical evidence of the value list's comprehensiveness. Moreover, the lack of a theoretical base rendered it difficult to recognize the implications of high priorities on one value type compared to other value types.

Building upon the importance ratings of Rokeach's (1973) value lists, Schwartz and Bilsky (1987) proposed a theory of the structure of human values. Based on earlier definitions of values, five important features of these definitions were extracted. Values were proposed to be concepts or beliefs about desirable end states or behaviours that transcend specific situations, guide selection or evaluation of behaviour and events, and are ordered by relative importance. Schwartz and Bilsky (1987) incorporated these features within their value theory, but extended the definition by focusing on the motivational and meaningful concerns embodied in values. It was postulated that values consist of universal cognitive representations of three basic human requirements: biological needs, social interactional requirements for interpersonal

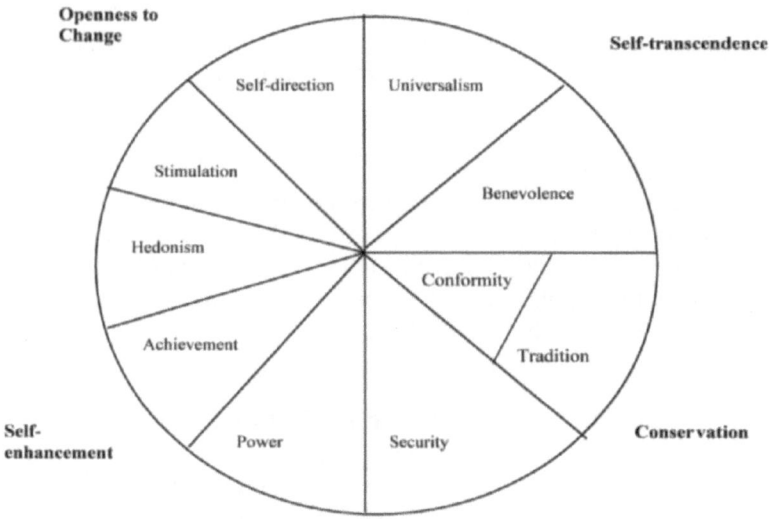

Source: adapted from Schwartz et al (2001)

Figure 5.1 *Theoretical model of structure of relations among*
ten value constructs

coordination, and social institutional demands for group welfare and survival. From these requirements a typology of values can be theorized. As seen in Figure 5.1, this typology can be divided into ten value types.

This structure of values refers to the conceptual organization of values based on their similarities and differences. Two different domains are conceptually distant if it is practically or logically contradictory to give high priority to values in both domains simultaneously. Two domains are conceptually close if placing high priority on values in both domains is compatible. This led to the assumption that people differ only in the relative importance that they place on a universally important set of value types. The implication of priorities on one value type for priorities on other types within the value system could accordingly be proposed. This value system can further be structured around two motivational dimensions, labelled 'openness to change versus conservation' and 'self-enhancement versus self-transcendence'. The first dimension adheres to the conflict between values in terms of 'the extent to which they motivate people to follow their own intellectual and emotional interests ... versus [preserving] the status quo and the certainty it provides in relationships with close others, institutions and traditions' (Schwartz, 1992). The second dimension concerns values 'in terms of the extent to which they motivate people to enhance their own personal interest (even at the expense of others) versus the extent to which they motivate people to transcend selfish concerns and promote the welfare of others, close and distant, and of nature' (Schwartz, 1992).

Value priorities concern guides for survival in the social environment, as well as principles for moral and ethical living. This view of values is reflected in Schwartz's (1999) definition of values as:

> *... conceptions of the desirable that guide the way social actors (e.g. organizational leaders, policy-makers, individual persons) select actions, evaluate people and events, and explain their actions and evaluations.*

The distinctiveness of the ten value types as proposed by Schwartz's value theory has found substantial support in samples of a wide range of different cultures (Schwartz, 1992, 1999; Schwartz et al, 2001) and has been assessed on a number of different issues – for instance, in work settings (Ros et al, 1999), out-group social contact (Sagiv and Schwartz, 1995), gender relations (Feather, 2004) and voting behaviour (Schwartz, 1996).

SOCIAL VALUE PRIORITIES

Personal value systems concern people's own judgements about their best possible living conditions, while social value systems concern their perceptions about the judgements of others. Personal value systems are linked to people's sense of self. Social value systems reflect interaction with other people or groups of people (Rohan, 2000). Although value theory recognizes the importance of social interaction for interpersonal coordination and social institutional demands for personal value priorities, there are reasons to believe that specific social value priorities might differ from personal priorities. Through interaction with others, new beliefs evolve. Such beliefs may be incorporated within the personal value system, but may also form a coherent system of social value priorities. To further clarify the distinction between personal and social value structures, both personal and social value systems are intrapsychic cognitive structures – not descriptions of groups' value systems. Since social value priorities are conceptions of others' judgements of the best possible living or functioning conditions, they organize people's perceptions of others. Moreover, if people's personal value structures are universal, then the perception of others' value priorities should also be organized according to this universal structure. Finally, if we accept that both personal and social value priorities exist, then attitudes and behaviour may be traced to either system (Rohan, 2000).

VALUES (AND NORMS) APPLIED TO ENVIRONMENTAL ISSUES

There have been several attempts to find relationships between values, on the one hand, and environmental attitudes or behaviour, on the other. A large body of research has identified environmental problems as the result of a tendency to act in one's own private interest (Hardin, 1968; Van Lange et al, 1992; Dawes and Messick, 2000). Values that could turn behaviour into a more environmentally benign direction would probably directly concern the environment or be values that give priority to things that go beyond one's own interest. In Schwartz's value theory, this corresponds to the self-transcendent value dimension. In one of the studies aimed at finding a relationship between values and pro-environmental behaviour, Karp (1996) measured self-reported pro-environmental behaviour together with the Schwartz 56-item value scale. Results showed that valuing the dimensions of self-transcendence and openness to change, and especially universalism and biospheric values, predicted pro-environmental behaviour, whereas the self-enhancement and conservation dimensions were found to be negative predictors.

Another study investigating the relationship between values and environmental behavioural intentions showed that self-transcendent values are predictive of people's self-reported willingness to take environmentally significant action. People with strong self-enhancement values were less supportive of pro-environmental actions (Stern et al, 1995). According to this study, the key to people's responses to any environmental problem was their values, as well as their specific beliefs about the consequences of environmental damage to the things they value. Respondents' beliefs about the adverse effects of environmental conditions also affected their willingness to act. These beliefs, however, were, in turn, affected by values. It was found that people with strong self-transcendent values were more likely to believe that environmental changes had adverse effects. These results were interpreted as indicating that values can affect pro-environmental action both directly and indirectly through beliefs about consequences (Stern et al, 1995). Values may affect these beliefs by simply motivating people to pay attention to information about how environmental conditions affect the things they value, or they may shape these beliefs directly by leading people to believe what they want to believe.

Using the Schwartz value scale, Schultz and Zelezny (1999) investigated cross-cultural relationships between values and attitudes in 14 countries. As a measure of environmental attitudes, a revised version of the New Environmental Paradigm (NEP) was used. The NEP scale is created to measure to what extent respondents consider humans an integral part of nature, which is similar to the concept of biocentrism (Stern and Dietz, 1994). The results showed a consistent pattern of relationships across countries. Universalism values predicted positive scores on the NEP scale, while the value types of power and tradition were negatively associated with NEP.

Environmental philosophers have often argued that there is a specific ecocentric ethic, based on ecocentric values, that assigns intrinsic value to the environment without concern for human welfare. For instance, Merchant (1992) makes a distinction between egocentric, homocentric and ecocentric ethics. This distinction is not explicit in Schwartz's (1992) value theory, in which environmental values can be regarded as either homocentric or ecocentric. To test if a specific ecocentric value structure can be discerned, Stern and Dietz (1994) selected items from the Schwartz's self-transcendence dimension to reflect biospherism and altruism, and added two biospheric items to be able to separate these two value orientations. Other value orientations measured were items from the self-enhancement cluster to reflect egoism, and value items from 'tradition' and 'openness to change'. Their hypothesis was that these value clusters are related to beliefs about environmental consequences (to the biosphere, to others and to the self) and to behavioural environmental intentions. The results failed to identify a distinct biospheric value orientation. As predicted by Schwartz's (1992) theory, the biospheric items could not be distinguished from the social altruistic items. Furthermore, the biospheric–altruistic value orientation was related to beliefs about the environmental consequences to the biosphere, to self and to others, as well as to behavioural intentions directly. Egoistic value orientation was found to be negatively related to beliefs about consequences to the biosphere and to behavioural intentions. Similar results were found in a later study (Stern et al, 1995) where subjects were found to construct attitudes to new or emergent environmental attitude objects by referring to personal values and beliefs about consequences to environmental conditions. Willingness to take pro-environmental action was found to be a function of both self-transcendent values and beliefs, with values also predicting beliefs.

Since altruistic motives are deemed important in expressing environmental concern, these results signify that environmental resources are considered public goods. This, in turn, denotes that psychological models of pro-social behaviour could be applicable in the environmental domain. Pro-environmental behaviour is seen here according to a moral perspective, with concern for future generations. Following this line of reasoning is research linking Schwartz's norm activation theory (Schwartz, 1977) to pro-environmental behaviour. This model for altruistic behaviour suggests that people recognize a particular situation as having adverse consequences for other people and that they feel a personal responsibility to take action to prevent these consequences. This theory has been extended to the environmental domain and used to predict several environmental behaviours, including yard burning (Van Liere and Dunlap, 1978), purchase of lead-free petrol (Heberlein and Black, 1976), energy-conserving behaviour (Black et al, 1985), recycling (Bratt, 1999; Hopper and Nielsen, 1991) and support for environmental protection (Stern et al, 1986).

Extending earlier research, a theory was presented that linked value theory, norm activation theory and the NEP scale through a causal chain leading to behaviour: the value–belief–norm (VBN) theory of environmental concern (Stern et al, 1999). Each variable is postulated to affect the next variable in the chain and may also directly affect variables further down the chain. This causal chain consists of personal values, the NEP, adverse consequences and ascribed responsibility beliefs, and personal norms for pro-environmental action. The theory, furthermore, postulates that the outcomes that matter in activating personal norms are adverse consequences to whatever an individual values. The theory was tested on different aspects of support for environmental movements: environmental citizenship, private-sphere behaviour and policy support (Stern et al, 1999). Of particular interest here are the relationships between values, norms and policy support. Stern and colleagues showed that people support policies when they subscribe to environmental values, perceive threats to these values, and feel a moral obligation to take pro-environmental action. Hence, to the extent that environmental values affect support for environmental movements, they do so sustained by personal norms. In technical terms, the effect of values on policy support is mediated by personal norms.

THE IMPACT OF VALUES ON ACCEPTING POLICY MEASURES

The immediate causes of global change are only partly a consequence of the choices that individuals make in their role as consumers. Consequently, the study of consumer behaviour can yield only partial knowledge about the causes of global change. Psychologists should therefore extend their focus by studying how organizational and governmental decision-making affect environmental resource use. Public organizations administer and implement policy tools to influence the behaviour of target groups and private citizens. It is important to understand how climate change policies affect, and are accepted by, private citizens, as well as companies and public administrations. In organizational settings, the set of norms, values, beliefs and goals that guide decision-makers most likely differ from what is of interest when they act as private citizens. To the extent that these organizational value systems are influential, value priorities that differ from priorities within personal value systems might affect subsequent norms and attitudes. In some instances, people must be willing to provide political support, while in others they must have the motivation and knowledge necessary to adopt new technologies. Thus, corporations make a greater direct contribution to environmental problems than individuals, and it is worth examining whether more can be done to alleviate these problems by modifying decision-making and behaviour in organizations.

The main question here is why some people or organizations see the benefits of policy measures and are prepared to accept their implementation, while others are less responsive. In the following two cases, we assume that to the extent that people care for the environment and believe that certain acts pose a threat to the environment, they are also likely to acknowledge an obligation to do something to avert these threats. Policy measures provide one means of achieving this. However, the extent to which people care is not assumed to be determined once and forever. Situational factors, such as which role one occupies, can affect the degree to which environmental considerations are examined.

HOW DO VALUES INFLUENCE ACCEPTANCE OF POLICY MEASURES IN ORGANIZATIONS?

The primary aim of the study by Nilsson et al (2004) was to examine how values, organizational goals and norms influence willingness to accept climate change policy measures within organizations. We expected self-transcendent values to guide acceptance through organizational norms. The mediating function of norms is based on the abstract and general characteristics of values, which are assumed to require a more specific representation in order to influence attitudes (i.e. acceptance). Another related issue was to test whether decision-makers in the public and private sectors emphasize different issues in supporting (or not supporting) policy measures, and if they vary with regard to values, norms, goals and acceptance.

A mail survey was administered to decision-makers whose organizations had a significant impact on greenhouse gas emissions in their region. Approximately half of the selected sample represented the public sector, working at the local government level. The other half represented the private sector – in particular, trade and industry. For the trade and industry sub-sample, companies with a significant impact on global climate change in terms of greenhouse gas emissions were selected. All organizations had at least 35 employees. Five groups of policy means were investigated: subsidies, taxes, information, regulations and emissions trading.[1]

We found that among decision-makers in the public sector, environmental values were important determinants of willingness to accept climate change policy instruments. This was true for all instruments except emissions trading. An unexpected result was that acceptance of emissions trading was not predicted by any of the proposed factors. In the public domain, it may have seemed unclear whether emissions trading would result in positive environmental consequences or not. In the private sphere, the lack of relationships implies that potential economic benefits for the companies were uncertain. We furthermore found the effects of self-transcendent values on acceptance of the policy instruments to be mediated by norms. These findings are in accordance with the hypotheses. However, for the private sector, no effects of values or norms were found. Rather, acceptance was predicted by internal goals (i.e. profit concerns). Self-transcendent values and external goals were more highly prioritized in the public sector, while self-enhancement values and internal goals were more highly prioritized in the private sector. Furthermore, decision-makers in the public sector held stronger norms and were more positive towards all policy instruments.

These findings indicate that among decision-makers in the public sector, self-transcendent values activate norm perceptions and a moral obligation to protect the environment. In the private sector, values and norms seem to have limited influence on attitudes towards investigated policy instruments. The implication is that organizations establish their own criteria for assessing policy options. These criteria are, in turn, guided by goals that organizations wish to promote. In the public sector, decision-makers attend to what they perceive as the 'will of the people'. To the extent that citizens care for the environment, decision-makers will respond positively. In the private sector, goals that promote self-enhancement values are more salient. As a consequence, self-transcendent values are likely to be suppressed.

A complementary account suggests that those who enter public organizations differ in their private value priorities compared to those entering private organizations. The difference originates in self-selection. If this were the case, one would expect that employees in the private sector, whether they respond as employees or as private citizens, would react in the same manner as to environmental policy measures. Our second case attests to this.

Do Values Influence Acceptance Among Decision-Makers in the Private Sector when Approached as Private Citizens?

In the study by Nilsson and Biel (2005), the primary aim was to examine how values and norms influence willingness to accept climate change policy measures among decision-makers in the private sector considered as private citizens. Following Rohan's (2000) distinction between personal and social value priorities, there may be a difference between what individuals as private citizen prioritize and what they prioritize as decision-makers in private companies. As seen from the results in the previous study, decision-makers in private organizations were not guided by environmental values when expressing their attitudes towards policy measures. Rather, goals such as profit maximization that promote the internal interests of the organization determined their attitudes. When addressed as private citizens, a private frame and value system could become more salient. Values of an increasingly self-transcendent nature are then more likely to spring to mind (compare Stern et al, 1999).

If environmental values and personal norms are more important when decision-makers are addressed as private citizens, it can also be presumed that attitudes differ when people are addressed within their organizations, compared to when they present themselves as private citizens. Together with previous research (Stern et al, 1999), this would also make a strong claim about the importance of values and norms as determinants of environmental policy measures among private citizens, in general.

To clarify this hypothesis, a new sample of respondents was selected according to the same criteria applied in the previous study. Hence, they were all decision-makers in companies in the trade and industry sectors that had at least 35 employees.[2]

Results showed that when people in business were approached as private citizens, the importance of self-transcendent values increased, while the importance of self-enhancement values decreased. Furthermore, obligations to prevent negative consequences from climate change were stronger when respondents were addressed as private citizens rather than as decision-makers in companies. Finally, acceptance of policy measures could be accounted for by environmental values and personal norms. Taken together, the results indicate that a personal value structure is evoked in the private domain, while a social structure is predominant in an organizational setting. Moreover, value priorities could differ between these two structures.

The Value of Nature

So far, we have considered values in line with how they are usually conceived within social sciences. There is a wide consensus that values refer to guiding principles for humans in society (see Reser and Bentrupperbäumer, 2005, for example). Values are also associated with something important and valuable.

Furthermore, values have moral and affective qualities and people may feel strongly about important values. It should also be clear that value is not equivalent to money (Dawes, 1988). The struggle for money can, however, be an indication of the importance of a value such as wealth (Biel and Dahlstrand, 2005). The next case investigates the evaluation of environmental resources by means of economic assessments. In particular, the concern is whether the evaluation of an environmental change through an economic measure captures the range of values that people might associate with environmental management. Before the case is presented, some background on the contingent valuation method is provided.

The contingent valuation (CV) method is based on welfare economics theory and has found extensive application in recent years. The premise behind CV is positive economic theory, which assumes that when confronted with a possible choice between two goods, economic agents have preferences for one over another. It also assumes that through its actions and choices, an economic agent attempts to maximize its overall level of satisfaction or utility (Mitchell and Carson, 1989). This implies that an individual has the knowledge to decide what he or she needs and also has a clearly defined rank order of value importance. As should be clear by now, this one-dimensional value concept as represented by welfare economics differs from the multidimensional classification of values as proposed by Schwartz (1992) and Stern and Dietz (1994) (see also Fischhoff, 1991; Gärling, 2002). CV is used in several countries in order to assess, among other things, the environmental effects of government policies. CV is also used to estimate existence values – for example, how much an individual values a wetland or a forest which he or she has no intention of visiting.

METHODOLOGICAL CONCERNS IN VALUATION STUDIES

One approach to acquiring a monetary estimate of natural resources is to ask how much people are willing to pay (WTP), at the most, for a collective asset. Another way is to ask how much they would require in compensation for a deterioration of a resource (willingness to accept, or WTA). Several measurement problems have been suggested with the use of CV methods. A large part of the problem with CV measures is that they are sometimes insensitive to the quantity or scope of the good provided. When subjects are asked to evaluate a good, their WTP equals their stated WTP when one or several other goods also are included. In other words, the subject is willing to pay the same amount for the smaller and the larger good if the former is embedded in the latter (Baron and Greene, 1996). This phenomenon is called the 'perfect embedding effect'. An ordering effect has also been found. It arises when subjects are asked their WTP for the smaller good just after they have been asked about a larger one. In such cases, they are only prepared to pay a much smaller amount for the smaller good than for the larger one. This demonstrates that a good seen as embedded within a larger good has reduced value (Baron and Greene, 1996; Baron, 1997). It has also been shown that when subjects are asked about their WTP for several environmental resources of the same kind, it is not much higher, if at all, than the WTP for just one of these resources. This has been termed the 'adding-up effect' (Baron and Greene, 1996).

Apparently, people do not seem to have a well-founded concept of the monetary value of environmental resources. Moreover, how much people are willing to pay does not capture basic value priorities (Fischhoff, 1991). Although values are part of people, people do not always know how to express them. To the extent that values are expressed, self-enhancement values seem to take precedence over self-transcendent or moral values (Guagnano et al, 1994). The following section examines whether willingness to accept measures provides a wider scope for expressing moral values.

Willingness to pay versus willingness to accept

A willingness to accept measure is provided by asking respondents about the minimum compensation that they would accept for a deterioration of a resource. Following standard economic theory, the WTA measure should give almost the same results as the WTP measure (Willig, 1976). However, an abundance of research has shown that WTA bids are, in fact, between two to four times higher than WTP bids (Knetsch, 2000). Despite the fact that it is acknowledged that WTA is, in many cases, the conceptually correct measure (US NOAA Panel, 1994), due mainly to difficulties in eliciting meaningful WTA responses, the use of WTP is still recommended.

The fact that WTP and WTA measures show a great deal of context dependence is not surprising. An identical good can actually be treated differently in different contexts. In the WTP condition, the context is gain, while in the WTA situation, the context is loss. The separate kinds of measures evoke different affective valuations that may have divergent effects on people's evaluations. More precisely, in the case of environmental valuation, the WTA measure could elucidate a moral transgression – that is, deterioration of a natural resource – in exchange for individual economic benefits. Thus, the trade-off that respondents are asked to make between money and natural resources becomes clearer and more explicit if they are asked how much they would want in compensation for the loss of the resource, rather than how much they would pay in order to prevent the loss. The latter question is easily understood as a contribution, and not as a valuation or trade-off. If this is the case, then people will report stronger moral and affective reactions when they are asked to state a WTA sum than when they are asked to state a WTP sum.

DOES WILLINGNESS TO ACCEPT ELICIT STRONGER EMOTIONAL AND MORAL REACTIONS THAN WILLINGNESS TO PAY?

To test the hypothesis that differences in feelings and moral perceptions can account for much of the observed WTA–WTP discrepancy typically observed for public goods, a simple real money dichotomous-choice experiment was set up (Biel et al, 2006). 99 students were randomly assigned to two groups of approximately equal size: one WTP group and one WTA group. The procedure of a single real money dichotomous-choice experiment was chosen because we wanted to make certain that the monetary choice conditions were identical in the two conditions. One week before the actual experiment, participants were contacted by telephone. A time for their participation in the experiment was agreed upon. They were also given one of two instructions. In the WTA group, participants were informed that they would receive 50 Swedish kronor (approximately US$7) for participating. In addition, they were informed that 100 Swedish kronor would be donated to the ongoing World Wide Fund for Nature (WWF) Protecting the Swedish Otter project. Participants in the WTP condition were only informed that they would receive 150 Swedish kronor for their participation. At the time of the experiment, participants were reminded about their compensation, and they were also informed that they had a choice. Those in the WTA condition were told that rather than donating to the WWF, they could keep all of the money for themselves. Participants in the WTP condition were informed that although they could keep the money, they could also split it and donate 100 Swedish kronor to the WWF and its otter project, while keeping 50 kronor for themselves.

The results support the hypothesis that the choice to donate to the WWF or not would differ substantially between the framings. In the WTP setting, only 9 out of 48 participants chose to donate, while in the WTA group, 23 out of 51 shared their compensation with the WWF. Thus, those in the WTA group were more likely to donate than participants in the WTP group. Furthermore, moral reactions, paired with affective responses, could account for the WTA–WTP disparity in donor behaviour. Participants in the WTA condition expressed stronger negative emotions, particularly shame, regarding not donating and perceived themselves as morally blameworthy in not donating. It appears that moral values attached to the environment are more easily expressed under a WTA framework.

SYNTHESIS AND IMPLICATIONS FOR POLICY IMPLEMENTATION

The three studies presented here investigated the relationship between environmental values and people's reactions to policy instruments. This knowledge may be used in future research and in political interventions aimed at changing or promoting these attitudes. Since it has been found that changing attitudes is more effective when the content of the message is adapted to the basis of the attitude (DeBono, 1987; Eagly and Chaiken, 1993), addressing people's values should be most effective if the attitude is used to express values. Should attitudes serve other functions – for example, economic consequences – addressing these concerns could be more efficient. Our studies indicate that private citizens and decision-makers in the public sector show the necessary prerequisites for addressing environmental values in order to promote acceptance. In the private sector however, attempts to promote acceptance by appealing to such values are likely to be futile.

A general finding was that the strength of this relationship varies across situations. Notably, it varies with how salient environmental values are. In general, salient values are more likely to influence evaluations and behaviour than non-salient values (see also Feather, 1995; Maio and Olson, 1995). What, then, determines the salience of environmental values? At the heart of the matter is whether the environment is regarded as a public or common good or as a private good. In the former case, self-transcendent values, including environmental values, and a moral perspective are more likely to come to mind. When the environment is perceived as a private good, self-enhancement values and thoughts about individual interest tend to be more prominent. Policy instruments can tip the balance one way or the other. It is normally understood that economic incentives will have some (strong or weak, but definitely positive) effect on actions. A number of studies (see Frey, 1997) have shown, however, that economic incentives may simultaneously crowd out moral motivation. The risk that policy instruments obscure important values is dependent upon the type of measure that policy-makers introduce. One example where a considerable out-crowding effect can be expected concerns tradable emission rights (Frey, 1999). The trade between prospective polluters results in a price for the right to pollute. The system thus effects a change in relative prices and induces a change in behaviour, now based on extrinsic rather than intrinsic motivation. The bargaining process between the buyers and sellers focuses their attention on the possibility of legitimately polluting the environment. The relative price determines their choice. Comments on Frey's article (Nyborg, 1999; Uusitalo, 1999) implied that among business firms, there might not be any intrinsic motivation or moral considerations in the first place. This could be the case; but the crowding-out effect could still be attributed to a 'higher' level. By introducing tradable emission rights, authorities signal that emissions which damage the environment need not be evaluated within the moral sphere.

The expressive implication for environmental taxes is somewhat different. Charging for negative environmental side effects makes it clear that this is an undesired activity that one should restrain from. Depending upon the situation, crowding-out effects should still be expected. However, emission taxes are likely to be less destructive to environmental morale. Moreover, since the relative price effect is equal in magnitude to tradable emission rights, taxes are likely to be more effective in protecting the environment. Likewise, smaller crowding-out effects should be expected when legal systems are used to influence behaviour. In this case, citizens receive signals that an environmentally friendly behaviour is expected, and even if the locus of control shifts away from the individual to a regulating body, private economic concerns are not the primary focus of attention. Rather, it is the protection of the public good that is at stake. Pointing to this hypothesis is the finding that all samples in the above studies (Nilsson et al, 2004; Nilsson and Biel, 2005) were more positive regarding legal rather than economic instruments. This difference was most notable in the private sector.

Evaluations and behaviour are also determined by whether people approach the situation with a focus on prevention or promotion (compare Higgins et al, 1994). On average, people seem to be more concerned about preventing negative environmental impacts than promoting positive environmental consequences (Grankvist et al, 2004). The fact that respondents were more inclined to donate to the public good in the willingness to accept than in the willingness to pay condition illustrates this phenomenon. The WTA framework evokes a loss context, an infringement on environmental values, while the WTP framework elicits a gain context.

Despite the fact that people honour environmental values, they may not do so in all domains. As for decision-makers in private companies, they were more likely to be guided by their environmental values when they evaluated policy instruments as private citizens than in their professional roles. Furthermore, decision-makers in the public sphere were more affected by environmental values and norms than their counterparts in the private sphere. Different domains foster divergent evaluative frameworks. To the extent that people identify themselves as part of a certain domain, they are also apt to behave according to value priorities within that domain. Should they identify themselves as belonging to another domain, their evaluative framework also changes.

Finally, the character of the resource may determine whether it is seen as a public good or not. Certain resources are more easily recognized as being at risk than others. Thus, most people can understand the fact that whales are running the risk of becoming extinct and that rain forests are disappearing at an alarming rate. It is more difficult to imagine that the air we breathe, or the climate we enjoy, is deteriorating. Once the risk is recognized, the common-pool resource dilemma presents itself and the question of resource management is also raised. A related phenomenon is the 'big-pool' illusion. Resources such as the air and sea may be perceived as endless. Only when people understand that they are scarce resources will they also comprehend that they are valuable common resources and that environmental as well as other values are affected.

The quality of the interaction that people have with the resource may also determine whether environmental or other considerations are raised. Turning back to the example of whales, economic values are likely to play a more prominent role among whale-hunting communities than among people not dependent upon whales for their income. While the latter group may be guided by biocentric values in applying an environmental valuation framework, the former group may be motivated by an efficient use of the resource in order to secure it for future generations. Thus, a central question for further research is when common resources are more likely to be perceived as private goods and when they will be seen as public goods. This, in turn, has consequences for which instruments should guide environmental policy. It should be kept in mind, though, that if people become used to valuing natural resources in terms of monetary incentives, there is a risk that self-transcendent values will retreat to the background and that the intrinsic value of nature will remain unnoticed.

NOTES

1   A mail survey was administered to selected decision-makers in the third largest region in Sweden (the Västra Götaland region, with approximately 1.5 million residents). The selected decision-makers were economically or environmentally responsible in their organizations.
2   The criteria for selecting the decision-makers were the same as in the previous study except that the sample was a national sample with the western region excluded.

REFERENCES

Baron, J. (1997) 'Biases in the quantitative measurement of values for public decisions', *Psychological Bulletin*, vol 122, pp72–78

Baron, J. and Greene, J. (1996) 'Determinants of insensitivity to quantity in valuation of public goods: Contribution, warm glow, budget constrains, availability, and prominence', *Journal of Experimental Psychology: Applied*, vol 2, pp107–125

Biel, A. and Dahlstrand, U. (2005) 'Values and habits: A dual-process model', in Krarup S. and Russell C. S. (eds) *Environment, Information and Consumer Behaviour*, Edward Elgar, Cheltenham, UK

Biel, A., Johansson-Stenman, O. and Nilsson, A. (2006) *Emotions, Morality and Public Goods: The WTA/WTP Disparity Revisited*, Working Papers in Economics no 193, Göteborg University, Department of Economics, Göteborg, Sweden

Black, J. S., Stern, P. C. and Elworth, J. T. (1985) 'Personal and contextual influences on household energy adaptations', *Journal of Applied Psychology*, vol 70, pp3–21

Bratt, C. (1999) 'The impact of norms and assumed consequences on recycling behaviour', *Environment and Behavior*, vol 31, pp630–656

Dawes, R. M. (1988) *Rational Choice in an Uncertain World*, Harcourt Brace Jovanovich, San Diego, CA

Dawes, R. M. and Messick, D. M. (2000) 'Social dilemmas', *International Journal of Psychology*, vol 35, pp111–116

DeBono, K. G. (1987) 'Investigating the social-adjustive and value-expressive functions of attitudes: Implications for persuasion processes', *Journal of Personality and Social Psychology*, vol 52, pp279–287

Eagly, A. H. and Chaiken, S. (1993) *The Psychology of Attitudes*, Harcourt Brace and Company, Forth Worth, TX

Feather, N. T. (1995) 'Values, valence, and choice: The influence of values on the perceived attractiveness and choice of alternatives', *Journal of Personality and Social Psychology*, vol 68, pp1135–1151

Feather, N. T. (1996) 'Values, deservingness, and attitudes toward high achievers: Research on tall poppies', in Seligman, C., Olson, J. M. and Zanna, M. P. (eds) *The Psychology of Values: The Ontario Symposium, Vol 8*, Lawrence Erlbaum Associates, Hillsdale, NJ

Feather, N. T. (2004) 'Value correlates of ambivalent attitudes toward gender relations', *Personality and Social Psychology Bulletin*, vol 30, pp3–12

Fischhoff, B. (1991) 'Value elicitation: Is there anything in there?', *American Psychologist*, vol 46, pp835–847

Fransson, N. and Gärling, T. (1999) 'Environmental concern: Conceptual definitions, measurement methods, and research findings', *Journal of Environmental Psychology*, vol 19, pp369–382

Frey, B. S. (1997) *Not Just for the Money*, Edward Elgar, Cheltenham, UK

Frey, B. S. (1999) 'Morality and rationality in environmental policy', *Journal of Consumer Policy*, vol 22, pp395–417

Gärling, T. (2002) 'Mätning av allmänhetens betalningsvilja: Översikt av metodrelaterad forskning', in Spolander, K. (ed) *Rationalitet och etik i samhällsekonomisk analys och Nollvisionen*, Verket för Innovationssystem (VINNOVA)/Nationalföreningen för Trafiksäkerhet (NTF), Stockholm, Sweden

Grankvist, G., Dahlstrand, U. and Biel, A. (2004) 'The impact of environmental labelling on consumer preference: Negative vs positive labels', *Journal of Consumer Policy*, vol 27, pp213–230

Guagnano, G. A., Dietz, T. and Stern, P. C. (1994) 'Willingness to pay for public goods: A test of the contribution model', *Psychological Science*, vol 5, pp411–415

Hardin, G. (1968) 'The tragedy of the commons', *Science*, vol 162, pp1243–1248

Heberlein, T. A. and Black, J. S. (1976) 'Attitude specificity and the prediction of behavior in a field setting', *Journal of Personality and Social Psychology*, vol 33, pp474–479

Higgins, E. T., Roney, C. J. R., Crowe, E. and Hymes, C. (1994) 'Ideal versus ought predilections for approach and avoidance: Distinct self-regulatory systems', *Journal of Personality and Social Psychology*, vol 66, pp276–286

Hopper, J. R. and Nielsen, J. M. (1991) 'Recycling as altruistic behaviour: Normative and behavioral strategies to expand participation in a community recycling program', *Environment and Behavior*, vol 23, pp195–220

Karp, G. D. (1996) 'Values and their effect on pro-environmental behaviour', *Environment and Behavior*, vol 28, pp111–133

Kerr, N. L., Garst. J., Lewandowski, D. A. and Harris, S.E. (1997) 'That still, small voice: Commitment to cooperate as an internalized versus social norm', *Personality and Social Psychology Bulletin*, vol 23, pp1300–1311

Kluckhohn, C. K. M. (1951) 'Values and value orientations in the theory of action', in Parsons, T. and Sils, E. (eds) *Toward a General Theory of Action*, Harvard University Press, Cambridge, MA

Knetsch, J. L. (2000) 'Environmental valuations and standard theory: Behavioural findings, context dependence, and implications', in Tietenberg, T. and Folmer, H (eds) *The International Yearbook of Environmental and Resource Economics 2000/2001*, Edward Elgar, Cheltenham, UK

Maio, G. R. and Olson, J. M. (1995) 'Relations between values, attitudes, and behavioural intentions: The moderating role of attitude function', *Journal of Experimental Social Psychology*, vol 31, pp266–285

Merchant, C. (1992) *Radical Ecology: The Search for a Liveable World*, Routledge, London

Mitchell, R. C. and Carson, R. T. (eds) (1989) *Using Surveys to Value Public Goods: The Contingent Valuation Method*, Resources for the Future, Washington, DC

Nilsson, A. and Biel, A. (2005) 'Acceptance of climate change policy measures: Role framing and value guidance', unpublished manuscript

Nilsson, A., von Borgstede., C. and Biel, A. (2004) 'Willingness to accept climate change strategies: The effect of values and norms', *Journal of Environmental Psychology*, vol 24, pp267–277

Nordlund, A. M. and Garvill, J. (2002) 'Value structures behind proenviromental behaviour', *Environment and Behavior*, vol 34, pp740–756

Nyborg, K. (1999) 'Informational aspect of environmental policy deserves more attention: Comment on the paper by Frey', *Journal of Consumer Policy*, vol 22, pp419–429

Reser, J. P. and Bentrupperbäumer, J. M. (2005) 'What and where are environmental values? Assessing the impacts of current diversity of use of "environmental" and "World Heritage" values', *Journal of Environmental Psychology*, vol 25, pp125–146

Rohan, M. J. (2000) 'A rose by any name? The values construct', *Personality and Social Psychology Review*, vol 4, pp255–277

Rokeach, M. (1973) *The Nature of Human Values*, Free Press, New York

Ros, M., Schwartz, S. H. and Surkiss, S. (1999) 'Basic individual values, work values, and the meaning of work', *Applied Psychology: An International Review*, vol 48, pp49–71

Sagiv, L. and Schwarz, S. H. (1995) 'Value priorities and readiness for out-group social contact', *Journal of Personality and Social Psychology*, vol 69, pp437–448

Schultz, P. W. and Zelezny, L. (1999) 'Values as predictors of environmental attitudes: Evidence for consistency across 14 countries', *Journal of Environmental Psychology*, vol 19, pp255–265

Schwartz, S. H. (1977) 'Normative influences on altruism', in Berkowitz L. (ed) *Advances in Experimental Social Psychology*, New York, Academic Press, vol 10, pp221–279

Schwartz, S. H. (1992) 'Universals in the content and structure of values: Theoretical advances and empirical tests in 20 countries', in Zanna, M. (ed) *Advances in Experimental Social Psychology*, New York, Academic Press, vol 25, pp1–65

Schwartz, S. H. (1996) 'Value priorities and behavior: Applying a theory of integrated value systems', in Seligman, C., Olson, J. M. and Zanna, M. P. (eds) *The Psychology of Values: The Ontario Symposium*, Lawrence Erlbaum Associates, Hillsdale, NJ, vol 8, pp1–24

Schwartz, S. H. (1999) 'A theory of cultural values and some implications for work', *Applied Psychology: An International Review*, vol 48, pp23–47

Schwartz, S. H. and Bilsky, W. (1987) 'Toward a psychological structure of human values', *Journal of Personality and Social Psychology*, vol 53, pp550–562

Schwartz, S. H., Melech, G., Lechman, A., Burgess, S., Harris, M. and Owens, V. (2001) 'Extending the cross-cultural validity of the theory of basic human values with a different method of measurement', *Journal of Cross-Cultural Psychology*, vol 32, pp519–542

Stern, P. C. and Dietz, T. (1994) 'The value basis of environmental concern', *Journal of Social Issues*, vol 56, pp121–145

Stern, P. C., Dietz, T., Abel, T., Guagnano, G. A. and Kalof, L. (1999) 'A value–belief–norm theory of support for social movements: The case of environmentalism', *Human Ecology Review*, vol 6, pp81–97

Stern, P. C., Dietz, T. and Black, J. S. (1986) 'Support for environmental protection – the role of moral norms', *Population and Environment*, vol 8, pp204–222

Stern, P. C., Dietz, T., Kalof, L. and Guagnano, G. A. (1995) 'Values, beliefs, and pro-environmental action: Attitude formation toward emergent attitude objects', *Journal of Applied Social Psychology*, vol 25, pp1611–1638

US NOAA (National Oceanic Atmospheric Administration) Panel (1994) *Natural Resource Damage Assessments: Proposed Rules*, US Federal Register, 59, 7 January, www.access.gpo.gov/su_docs/aces/aaces002.html, accessed 8 May 2000

Uusitalo, L. (1999) 'Do the same models apply to consumers and business firms? Comments on the paper by Frey', *Journal of Consumer Policy*, vol 22, pp435–438

Van Lange, P. A. M., Liebrand, W. B. G., Messick, D. M. and Wilke, H. A. M. (1992) 'Introduction and literature review', in Liebrand W. B. G., Messick D. M. and Wilke H. A. M. (eds) *Social Dilemmas: Theoretical Issues and Research Findings*, Pergamon, Oxford, UK

Van Liere, K. D and Dunlap, R. E. (1978) 'Moral norms and environmental behavior: An application of Schwartz's norm activation model to yard burning', *Journal of Applied Social Psychology*, vol 8, pp174–188

Willig, R. (1976) 'Consumer's surplus without apology', *American Economic Review*, vol 66, pp589–597

6

Organizational Culture, Professional Norms and Local Implementation of National Climate Policy

*Chris von Borgstede, Mathias Zannakis and
Lennart J. Lundqvist*

INTRODUCTION

We have seen that Sweden's central government opted for an implementation of the national climate strategy, which expects 'each and everyone' to take appropriate action. But such multilevel and multi-actor governance will encounter constraints that are strongly embedded in the present structure and culture of public affairs. We find local jurisdictions that run counter to an ideal organization for effective climate policy implementation (see Hooghe and Marks, 2003). Swedish local government is based on 'territory', charged with a multitude of sometimes competing responsibilities organized along 'sectoral' lines, and with long-term legitimacy. Responsibilities for such issues as transport, energy, physical planning, housing and the handling of material flows through society are either explicitly placed on local governments or constitute legally binding regulations and constraints on municipal action (Cabinet Bill 2001/02:55). This potential clash between effectiveness and legitimacy comes out clearly in the Climate Commission's final report (SOU 2000:23, p242):

> ... *it is probable that both already implemented organizational reforms and the present administrative structure are sub-optimal for an effective implementation of the climate policy. The diffusion*

> *of responsibilities among a large number of agents makes it diffi-
> cult to get an overview of measures, results and lines of
> responsibility.*

However, such structural contexts do not in and of themselves make or break climate policy. Our basic assumption here is that they affect the norms and notions of climate change 'problems' and 'solutions' held by local decision-makers and by affected interests and targeted actors. This chapter therefore specifically analyses the motivations of local policy-makers, administrative officials and private business actors when making decisions on climate-related issues. We ask how organizational norms and 'cultures', as well as professional norms, actually influence decision-makers' views on:

1   the seriousness and relevance of climate change for society and its organizations;
2   the distribution of duties and responsibilities for acting on climate issues;
3   different policy instruments directed at mitigating climate change; and
4   actual collaboration in broader programmes/projects related to climate change.

This chapter specifically addresses the third and fourth of these issues. It does so by presenting the results from two different analyses of a survey of local private and public decision-makers in the Västra Götaland region of Sweden, as well as the results from an analysis of a series of semi-structured follow-up interviews with decision-makers in four selected local governments in the Greater Gothenburg area of Sweden.[1]

SOCIAL DILEMMAS, NORMS AND THE READINESS FOR COOPERATION IN LOCAL CLIMATE POLICY IMPLEMENTATION

Climate change does not respect political, administrative or organizational boundaries, but accentuates the interdependence of these units. Therefore, it poses a social dilemma to decisions-makers: should we try to initiate cooperation to combat climate change, even if this implies costs to us that we might not be able to recover while non-participants might benefit without contributing time and resources? Climate change forces individuals or groups to choose whether to contribute or not to a common resource, and to harvest as much as they want or to limit their use of the resource (see Messick and Brewer, 1983; Dawes and Messick, 2000). An example is the choice facing urban commuters: should we use our own car or opt for public transportation? Most individuals state that their first preference is to use their private car for reasons of (immediate) comfort and flexibility. Yet, if all were to follow this preference, the result would be increases in carbon dioxide (CO_2) emissions, leaving everyone suffering from the collective effects of individual choice (see, for example, Van Vugt et al, 1996; Nordlund and Garvill, 2003).

However, pro-environmental behaviour and readiness to cooperate – that is, to act in favour of the collective, rather than just pursuing one's own interests – are not as rare as we would expect. Social and personal norms can trigger cooperative behaviour among key implementation actors (Vining and Ebreo, 1992; Nilsson et al, 2004). Our analysis links cooperative behaviour to moral considerations (norms) and organizational culture. The structural contexts of local governance affect local decision-makers' norms and notions of climate change 'problems' and 'solutions'. Over time, this creates quite distinctive intra-organizational 'cultures' that foster assumptions, beliefs and norms about what 'is' and what 'should be [done]'(Alvesson, 2002). Views on agreeable alternatives and solutions to climate change may thus differ among sectoral units in a local government organization (see, for example, Newman and Nutley, 2003). At the same time, there are distinct professional 'communities of practice' sharing a specific view of a problem and its solution (Brown and Duguid, 1991). Decision-makers with professional norms differing from those predominant in the organization may choose not to accept decisions to implement the national climate strategy if they see this as against their professional convictions (see Selden et al, 1999) or as involving a loss of professional esteem and respect (Haslam et al, 2000).

In a broader perspective of local governance, it is reasonable to assume that public and private organizations differ in objectives and norms. Whereas the private sector is generally more guided by values of profit maximization, local authorities are mandated to promote norms and objectives linked to the 'common good'. Admittedly, this 'common good' might be differently – and even contradictorily – interpreted among different sectors of local government. However, we still assume that the public sector is generally more inclined to choose cooperation – that is, to contribute to the solution of climate issues – than is the private sector. This will affect the degree of actual public and private involvement in climate-related collaboration projects. Given the existence of distinct professional norms, we furthermore assume that environmental professions might be more ready to cooperate on climate problems than other professionals in both the public and private sectors.

However, one should not forget that other factors than norms might influence cooperation on climate policy implementation. Resources are important; lack of resources can be an obstacle. Readily available resources will strengthen the impact of norms on involvement in collaboration projects, whereas lack of resources will reduce their impact. Historic behavioural patterns are another factor. The scope and intensity of 'institutionalized' conflicts affect cooperative behaviour. Furthermore, actors' perceptions of the clarity of the political or managerial guidelines for an issue area might affect cooperation.

CULTURE, NORMS AND THE ACCEPTANCE OF SPECIFIC CLIMATE POLICY MEASURES

Our major assumption has been that distinct and, perhaps, even rival forms of organizational norms[2] fostered by different local structures and units will affect the implementation of national policy instruments. The 'carrots, sticks and sermons' in the national climate policy may be deemed more or less compatible with the organizational or individual norms already held in different parts of the local governance structure. Some instruments may be seen as associated with low levels of sacrifice and little inconvenience, while others seem linked to high costs and much inconvenience.

We therefore distinguish among different climate policy measures in line with the *low-cost hypothesis*. This relates the costs and sacrifices involved in decision-making situations to the ease or difficulty that local actors will have in transforming environmental concern into behaviour correspondent to the policy instrument. If they perceive the costs and sacrifices implied by the instrument as too high, environmental concern does not suffice to make them overcome their reservations. Pro-environmental attitudes or general environmental concern, therefore, will have little effect and implementation will suffer (Diekman and Preisendörfer, 2003).

The argument for seeing perceptions of low cost and high cost as relevant in predicting actors' viewpoints on climate policy instruments is thus as follows. With low costs and no obvious sacrifice, an actor more easily transforms his or her attitudes and norms into the behaviour induced by that instrument. A more easily forthcoming acceptance of national policy instruments and related organizational changes can thus be expected in situations where instruments are perceived as carrying low costs and few sacrifices. The opposite can be expected in situations with high costs and severe inconveniences (Diekman and Preisendörfer, 2003). We assume that this may be more pronounced if the actor also has a high concern for the environment.

In sum, organizations and individuals at the local level will differ in their perceptions of national policy instruments and measures in terms of costs and inconveniences. Individuals working with the long-term protection of natural resources already share organizational and professional norms and practices compatible with the national climate strategy. In organizational units responsible for environmental control, acceptance of the national package of measures may therefore come more easily (see, for example, Bruff and Wood, 2000; Tait and Campbell, 2000). On the other hand, decision-makers in units dealing with infrastructure and collective transport systems, or housing and energy, may view climate policy measures as involving clashes with deeply held professional norms about desirable and appropriate measures for socio-economic development. The professional planners constitute an interesting key group. Engaged as they are in the welfare and further development of the municipality, they may interpret climate policy measures as another constraint on the local planning monopoly enjoyed by municipal governments. At the same time, the longer-term perspectives of local physical planning may lead them to an assessment of costs and sacrifices more favourable to effective local climate policy implementation.

Elected local politicians are responsible for enhancing the local 'common good' within their geographically delineated municipality. Furthermore, they are dependent upon the next election for their positions. They may view long-term climate policy measures as yet another intrusion on local self-government for different but (in the case of climate policy implementation) converging reasons. First, they may not conceive the local climate situation as particularly troublesome. This could induce them to reason in ways typical for social dilemma situations: 'Why should this local government engage in action towards climate change when problems do not originate here, when the effects of our action are, at best, infinitesimal, and when they might benefit other local governments that do not contribute to abatement actions?'

Second, the inherent multilevel governance character of climate policy adds to local hesitance. The distribution of political and administrative authority is such that an individual local government is never the exclusive 'owner' of problems contributing to greenhouse gas emissions. A typical example is the major national traffic arteries passing through local communities. They are a national responsibility. The same holds for the technical emission performance standards of the vehicles passing along the highways. Only the largest cities in Sweden have used the possibility of passing local 'environmental zone' ordinances to regulate what types of lorries and vans may be allowed where, if at all, in the inner city (SFS 1998:1276, Chapter 10, 1§ pt 5; see, for example, Skagersjö, 2002).

Another example concerns large-scale facilities that emit greenhouse gases. Municipalities are just one party in the court-like procedures for deciding if and under what conditions such facilities should be permitted to start up or continue their activities. Municipal environmental inspectors cannot introduce demands that exceed what is stipulated in the environmental courts' decisions (SFS 1998:808, Chapter 20). One should also recognize the impact of global emissions trading on local governments' ability to take action against climate change. Trading quotas are decided on a national level. Large companies emitting greenhouse gases are not required to negotiate with, or seek recognition of their trading activities from, municipal authorities (SFS 2004:657). Thus, local politicians and administrators may ask themselves why they should care for a problem that they do not 'own' – that is, for a problem which they lack the authority to tackle by themselves (see also Chapter 7 in this volume).

LOCAL CLIMATE COOPERATION: FINDINGS AND IMPLICATIONS FOR CLIMATE STRATEGY IMPLEMENTATION

Given that organizations differ in cultures and objectives, we assume that the degree of involvement in climate collaboration projects differs between the public and private sectors. The actual impact that organizational norms will have on involvement in collaboration projects is hypothetically mediated by available resources, intra-institutional obstacles and perceptions of political guidelines. These factors can either strengthen or weaken the impact of norms on involvement in specific climate collaboration projects (the following builds on von Borgstede and Zannakis, 2005).

We asked our respondents to specify if they engage in such collaboration projects on climate issues and, if so, what kind of projects they have been involved in (business-related activities, informational campaigns, education, technical development, behavioural changes, promoting energy efficiency, social planning, pro-environmental purchasing).[3] The general pattern confirms our expectations. While three out of out of five surveyed public organizations (62 per cent) were involved in some climate-related collaboration projects, only two out of five (43 per cent) among private organizations reported any such cooperation.

We found a positive correlation between organizational norms[4] and degree of cooperation for both the public and the private sector. Neither respondents' views on resources and intra-institutional obstacles, nor their views of the clarity of political guidelines seemed to have any influence on the degree of cooperation. Furthermore, views about the organization's own responsibility seem more strongly correlated with degree of collaboration for the public sector than for the private sector.[5] Finally, views of available resources are strongly correlated with (un)willingness to cooperate in the private sector, but not with the public sector's readiness for cooperation. These sectoral patterns also remain after checks of intervening or mediating effects from intra-institutional obstacles and political guidelines. Thus, what explains public and private organizations' involvement in climate cooperation projects are quite different factors.

Our results carry some quite interesting implications for the possibility of implementing a national climate policy effectively. The differences in degree of involvement in local cooperation projects between the public and private sectors have a long history. Local administrations have an institutionalized and legitimate role in inducing cooperation projects with key local interests and actors. This means that local public administrations have a longer tradition of collaborating within local government and with the private sector. They have, in short, a central role in creating local spheres of governance.

The difference in actual behaviour between the public and private sectors is clearly troublesome in terms of the auspices for effective climate policy implementation. The fact that prescriptive organizational norms (what one ought to do) do not seem to encourage cooperative readiness for the private sector can be seen as an indication that involvement in cooperation projects depends heavily upon traditional institutionalized organizational cultures spelling out what the organization can and cannot do. When a 'new' issue such as climate change appears on the agenda, organizational actors may be leaning back on engrained logics of appropriateness, rather than jumping into the unknown where they cannot calculate the 'return expected from alternative choices' (see March and Olsen, 1989).

Thus, when climate change policies and collaboration projects related to policy implementation are seen as a political novelty, the modest degree of involvement in collaboration projects from the private sector may not be too alarming from the viewpoint of effective policy implementation. National climate policy claims on local action should be seen as challenges, demanding the development of new organizational norms of what should be done and routines for what can appropriately be achieved (see March and Olsen, 1989; Powell and DiMaggio, 1991).

Nevertheless, the difference in actual behaviour between the public and private sectors is clearly troublesome for an effective climate policy implementation based on consensus. The factors brought up in the study bring forth the tension between self-interested and altruistic behaviour. The implementation of climate change policies thus equals a social dilemma situation. Our results indicate that there is no easy solution to this dilemma. It is notable that neither public nor private respondents seem prepared to assign an important role to their own organization in collaboration projects. Measures directed towards the private sector's norms may not necessarily lead to increased climate collaboration as long as available resources and economic incentives are much more crucial for private involvement in such projects.

LOCAL ACCEPTANCE OF CLIMATE POLICY INSTRUMENTS: FINDINGS AND IMPLICATIONS FOR CLIMATE STRATEGY IMPLEMENTATION

Climate policy measures differ with respect to the latitude of choice provided to local and individual actors. Car owners cannot escape paying carbon dioxide (CO_2) taxes every time they fill up the tank, but they can choose whether or not to heed climate information campaigns to leave the car and use collective means of transportation (see, for example, Chapter 4 in this volume). In the same way, local governments can choose whether or not to initiate collaboration projects to increase their chances of getting state subsidies for climate investment programmes (see Chapter 2 in this volume), or to accept wind farms to increase the nation's share of climate-friendly energy production (see Chapter 7 in this volume).

Still, it is evident that local and individual acceptance and 'legitimization' of climate policy instruments are crucial for an effective implementation of the national strategy. As outlined earlier, we have characterized climate policy measures in terms of 'low cost' or 'high cost' for the individual's organization and/or for the groups targeted by those instruments.[6] At issue, then, is whether there are patterns of acceptance/non-acceptance among local actors connected to organizational norms and/or individual (professional) role conceptions. To repeat our question: do the viewpoints of local actors on the acceptability of policy instruments depend upon where they 'sit' or upon what they 'do'?

Our analyses clearly show a significant difference in willingness to accept low-cost compared to high-cost strategies (the following builds on von Borgstede and Lundqvist, 2006). As expected, actors in both public and private organizations were significantly more willing to accept low-cost than high-cost strategies. From there, however, public and private organizations seem to part company. Respondents in public organizations were significantly more ready than those in the private sector to accept *all* climate policy strategies, both low- and high-cost ones. Where one 'sits' – that is, what predominant organizational norms one is exposed to – thus seems to moderate the degree of acceptance of climate policy measures. Low-cost measures would thus be more easily accepted, and effectively implemented, regardless of what sector or target group they are intended to affect.

This should not, however, be taken to mean that organizational affiliation excludes influences on the acceptance of policy instruments from individual professional roles – that is, what one 'does' within the organization. We found distinct patterns in acceptance of policy measures among environmentalists, planners and economists in the public and private sectors. Not surprisingly, all three groups rate low-cost strategies as more acceptable than high-cost ones. However, regardless of organizational affiliation, environmentalists stand out from the other two professions. First, they are significantly more willing than other planners and economists to accept high-cost strategies. Second, they are also more willing to accept high-cost measures than those who work with planning or economic issues. Evidently, regardless of one's organizational affiliation, what one 'does' in an organization does have an impact on willingness to accept climate policy measures.

The actors' degree of concern for the environment might affect their acceptance of climate policy measures.[7] We tested this in a series of analyses, together with the importance of organizational and societal norms (for details, see von Borgstede and Lundqvist, 2006). In the private sector, environmental concern and norms were unrelated to acceptance of high-cost strategies. However, in the public sphere, environmental concern was positively related to acceptance of these strategies. This implies that environmental concern paves the way for acceptance of high-cost strategies in the public, but not in the private, sector. Since neither organizational nor societal norms had an impact on acceptance, social forces to accept high-cost strategies are absent in either sector. As for low-cost strategies, both environmental concern and social norms had a positive effect on acceptance in both sectors. To the extent that one perceives climate change as a genuine threat, and /or believes that our national government should take action towards climate change, one is also likely to accept low-cost strategies.

Our analysis of organizational affiliation, professional roles and acceptance of climate policy instruments thus yielded some systematic patterns. An environmental professional role tends to lead to a higher level of acceptance of both low-cost and high-cost policy measures than those found for planners and economists. We discuss the implications of these findings for the implementation of the national climate strategy along three lines:

1 using low-cost climate policy measures (information, subsidies and regulations);
2 changing the institutional context of local climate policy; and
3 solving conflicts over the distribution of responsibility.

Earlier analyses suggest that environmental concern is crucial to more environmentally friendly action in low-cost situations. However, we found that environmental concern has an impact on the acceptance of both low-cost and high-cost measures. We see two partially contradicting implications of this for a more intensive use of low-cost climate policy measures. One is the low-cost theory prediction that environmental attitudes provide more important normative guidance to act in situations that do not demand extensive changes in attitudes or behaviour. This should not be taken to mean that pro-environmental concern plays a role only for very narrow aspects in environmental protection. The assumption should, rather, be that it is more likely for pro-environmental individuals than for those not embracing environmental concerns to give preference to environmentally friendly alternatives in situations of choice.

At the same time, however, a heightened awareness of *global* climate change brought about by low-cost information measures should not be taken as automatically leading people at the *local* level to adopt more climate-friendly behaviour – to link their own behaviour in the local domain to negative effects on the climate. A recent study indicates that people tend to view global environmental problems as more serious than regional or local ones. People also tend to assign least responsibility for global environmental problems to individuals, and less responsibility to local organizations than to national and international levels (Uzzell, 2000).[8] Nevertheless, we contend that existing differential perceptions of responsibility for local–global environmental problems must be taken into account when designing climate policy measures. One way could be to link the effects of local socio-economic development to greenhouse gas (GHG) emission trends. This could make low-cost information measures more effective in bringing about such norm changes across local organizational cultures that are conducive to successful local climate policy implementation.

Organizational cultures are strongly entrenched in the institutional context of local politics. Sweden's municipal governments enjoy constitutional and legal powers to determine future physical and socio-economic development in their community. Local councillors are largely elected based on their policies and plans for short- to medium-term economic and social welfare developments. They may thus see the implementation of such plans as politically more rewarding than adapting to nationally imposed climate measures whose positive effects are visible only in the very long run. Planning, infrastructure and economic departments have a strong position in strategic local governmental decision-making, which implies that nationally imposed climate policies and local plans for socio-economic development may not easily square with each other. This is even more so as local developmental action (infrastructure, housing, business districts, etc.) can be defended as nothing but a 'fly in the cathedral' – that is, as having an infinitesimal impact on global climate change. However well-motivated changes in local governmental powers would be for an effective implementation of climate policy measures, redrawing the lines of authority and responsibility might be too 'high cost' in terms of legitimacy to rank as a feasible alternative.

RESPONSIBILITY, COMPETENCE AND PROBLEM OWNERSHIP: THE LOCAL DILEMMA OF CLIMATE STRATEGY IMPLEMENTATION

The common-pool resource character of climate, and the existing distribution of authority and responsibility, create intricate problems of assigning – as well as of accepting and being able to carry out – the responsibility for climate change abatement. Greenhouse gas emissions are both international and local in origin, and they have internationally dispersed, as well as locally concentrated, effects. Finding a rational and, at the same time, politically feasible distribution of responsibility for abatement action is thus an extremely delicate affair. A purely geographical approach, whether it is based on local or national jurisdictions, may lead both those governmental levels to argue that not all of the abatement is 'their table'. Sectoral approaches involving private producers may lead to protests against 'unfair or uncertified loads' of responsibility. Discussions of consumer- and producer-based approaches to ascribing responsibility for climate change problems point to further intricacies (see Bastianoni et al, 2004).

In our interviews with local councillors and administrators in four strategically selected local governments in the Gothenburg region,[9] we approached this issue in a twofold way (see von Borgstede and Lundqvist, 2007). First, we asked about their recognition of climate as a collective resource, and about their recognition of possible conflicts between local governmental action and climate change abatement. Do they see a social dilemma in that their measures, however well intended from a local point of view, may, indeed, counteract national or global conceptions of, and strategies for, climate as a sustainable collective resource? Second, we asked about their views of municipal responsibility and competence (vested in legal authority) in order to capture the local actors' views on 'problem ownership'. What can local governments actually do about climate change, given current distributions of political and administrative authority and responsibility for sectors generating, or being affected by, climate change?

A key factor in assessing, acknowledging and accepting responsibility is whether local governments have recognized climate change as a relevant part of their political agenda. The general impression from our interviews is that virtually none of the respondents acknowledged that their local government had officially recognized climate change as a central item on the political agenda. Administrative discussions on problems related to climate change tend to treat such problems as 'sectoral' issues, such as waste management, energy consumption and water supply. Our respondents attributed this lack of specific attention to *climate* problems to perceived ignorance and lack of understanding. As for the small island municipality, interviewed actors said that the municipality does not contribute to climate change; it is simply not 'their table'. In Gothenburg's city government, the pattern is somewhat mixed. Asked whether they had paid any attention to the climate issue, three out of four answered that climate issues come in as one theme in the 'catch-all' view of environmental issues on the local agenda.

What is evident here is the importance of municipal size. The only one among selected municipalities that fully identified itself as a problem owner with respect to climate change was Gothenburg City. Both councillors and administrators recognized that the city is the source of a large amount of greenhouse gas emissions. The city harbours several heavy industrial facilities, such as the oil refineries that emit greenhouse gases. The size of the city and its function as the largest seaport in Scandinavia generates substantial traffic. Thus, the activities within the municipality constitute a source of problem ownership. The councillors and administrators furthermore recognized that municipal size and administrative capacity are 'symbiotic'; the city administration simply has so much expertise in the different sectors that it can take a lead in tackling many of the climate problems originating in the region.

Gothenburg councillors and administrators thus defined their municipality as a major originator of GHG emissions and the local government as having the capacity to influence and shape climate-related activities in line with the ambitions of the national climate strategy. In the other three municipalities, our interviewees invariably came up with road traffic as a very serious threat to climate. At the same time, they recognized that local government has little or no authority to regulate traffic streams on major traffic arteries, not to mention regulating the environmental performance of road vehicles.

How are these views on problem ownership related to local decision-makers' recognition of climate change as a social dilemma? Our interviews reveal that a majority of the respondents spontaneously think about climate problems and their own municipality's activities as a social dilemma situation:

> ... obviously there is a huge conflict ... in the way decision-makers in the municipality always try to distinguish the environmental questions [between] local, regional and global questions, and the further from the local level (which climate change actually is) the harder it is to get people to accept changing their current behaviour.

> ... of course, there is a conflict between own interests and the best for the collective as such; that's why we need to create motivation and knowledge.

In terms of which instruments to use in trying to solve this conflict, respondents first and foremost pointed to stricter laws and regulations as the major alternatives. Some also emphasized the need for more information to bring the climate change issue onto the local agenda, to increase knowledge about the origins, character and effects of climate change, as well as to create motivation for behavioural change. In order to provide the full picture, one should note that some respondents, particularly in the smaller municipalities, admitted that they had never thought about climate change in these terms. These actors, in addition, strongly argued against further national regulations to change people's lifestyles and behaviour.

CONCLUSIONS: NORMS, DILEMMAS AND POSSIBILITIES OF EFFECTIVE AND LEGITIMATE MULTILEVEL GOVERNANCE OF CLIMATE CHANGE

Using a unique set of local survey and interview data, we have analysed three aspects crucial to the local governance of climate change:

1 readiness to engage in collaboration on climate-related projects;
2 acceptance of different climate policy instruments; and
3 views on the distribution of duties and responsibilities to act on climate issues.

We worked from an explicitly stated premise in Sweden's national climate strategy: that the local level is of utmost importance for successful multilevel governance of climate as a collective resource. Furthermore, we have worked from the assumption that making decisions related to climate change poses a social dilemma to individuals and organizational units at the local level.

Our results have some important implications for a strategy of voluntary local collaboration in climate policy implementation. Decision-makers in the private sector do not see involvement in such projects as central to their organizations. We found that there is no significant difference between the public and private sectors' support for pro-environmental norms. However, these norms seem to affect only the public sector's readiness to engage in collaboration projects on climate issues. Resources for dealing with these issues are clearly important. Private actors tend to view resources as scarce, which means that pro-environmental behaviour is less forthcoming. Actors facing a social dilemma situation and focusing on resources and economic outcomes, rather than on social or moral normative arguments, are more inclined to recognize individual consequences and are less responsive to pro-social choices (compare Pillutla and Chen, 1999; Biel, 2003).

Simply put, local public-sector units seem normatively prepared to engage in local climate cooperation, while the private side is hesitant, sometimes even reluctant, on the grounds of scarce resources. For a policy based on multilevel governance as a major vehicle of implementation, this implies that means must be found to involve all affected interests in the implemention process. The inclusion of private organizations in collaboration projects as a complement to traditional public implementation of policy measures must, so runs our argument, proceed from knowledge of what factors trigger private involvement in collaboration projects for the sake of the common good.

Our findings point to the possibility of combining different low-cost policy measures – information, regulations and subsidies – in order to prevent climate policy implementation from stumbling on normative or institutional blocks down the local road. Information and regulatory changes, backed by state economic support, provide a viable alternative for lowering the costs and gaining private-sector acceptance. The historic experiences from social welfare implementation provide a prime example here. Our results thus point to some ways out of social dilemmas in local climate policy implementation without having to take recourse to such draconian high-cost policy measures, such as comprehensive restrictions on local governmental authority, and/or stiff regulations of private behaviour.

This brings us to a discussion of how to proceed in ascribing responsibility for climate change action to different actors or levels in multilevel climate governance. The further clarification of relationships among organizational and societal norms, environmental concern and allocation of responsibility should proceed from the idea that the attribution of responsibility is based on the social desirability of certain behaviours (see Devos-Comby and Devos, 2001). Norm conflicts among organizations as well as professions may arise when responsibility is assigned by one actor to other actors because he or she sees these actors as acting contrary to the first actor's norms. This becomes even more important to recognize as national climate measures are imposed on local actors with such different normative predispositions as the ones we found in our interviews (von Borgstede and Lundqvist, 2007).

The combined evidence from this study is that environmental concern and societal norms could make actors who are crucial to effective implementation more willing to accept climate policy measures. It is, however, only one step forward. More research is needed on what brings actors with different degrees of environmental concern and different normative orientations to accept cooperation as a way of meeting expectations imposed from above. Until then, national efforts to get climate change measures implemented by local governments who enjoy constitutionally guaranteed spheres of authority should proceed with a due amount of caution.

NOTES

1   The answers were sought in a two-stage process, involving a postal survey and 19 semi-structured follow-up face-to-face interviews. The survey took place during autumn 2002 with a target population of 756 selected decision-makers in the Västra Götaland region in Sweden. The response rate was 51 per cent, and we base the analyses on 356 usable questionnaires. Our respondents were recruited according to four criteria. First, they should be decision-makers within organizational units with an actual or potential climate impact. Half of the sample worked within local government (public sector) and the other half were decision-makers in trade and industry (private sector). Second, respondents should work mainly with either economic or environmental issues within their organizational unit regardless of sector (public or private). Third, respondents in the private-sector sub-sample should work in companies with an impact on climate change – that is, companies within the energy, transportation, oil, and building and construction sectors. Finally, only organizations with at least 35 employees were sampled in order to make sure that both economic and environmental decision-makers would be included. See also note 8, below.

2   Our survey contained items intended to measure two categories of *norms*:
    i   respondents' views of what their own organization should/should not do regarding climate change (here called *organizational* norms); and
    ii   respondents' views of what the national government should do regarding climate change (here called *societal* norms).
    Four items captured organizational norms, and three items captured societal norms. *Responsibility* for taking action regarding climate change was measured by the following assertion: 'It is not our organizations' responsibility to deal with climate change.' Furthermore, the moral dedication was considered from an organizational point of view – for example, beliefs that 'Within our organization we are prepared to put more effort on behalf of climate change issues.' Each of the normative belief items were rated on a five-point Likert scale ranging from (1) strongly disagree to (5) strongly agree.

3   Respondents were not limited to specifying one alternative. This resulted in a scale from (0) no kind of involvement in collaboration projects to (8) all possible

alternatives. Respondents were also asked to estimate how clear the *political guidelines* are from the municipality, the Västra Götaland region, and the state. Organizational resources were measured by asking respondents to indicate to what extent they consider their own organization's resources (time, personnel, knowledge and material resources) to be appropriate in tackling issues related to climate change. *Intra-institutional obstacles* were measured by asking respondents to indicate to what extent they consider it difficult or easy to:

- work with climate-related issues;
- cooperate with other relevant organizations about climate issues;
- cooperate with different units to facilitate the organization's work with climate issues; and
- devise strategies to reduce the organization's impact on the climate.

Answers could be given on four- or five-point Likert scales.

4   See note 2, above.

5   Again, see note 2.

6   *Climate policy measures* were typologized into five items: subsidies, regulations, taxes, information and emissions trading. With the exception of emissions trading, each of these measures was exemplified within areas of transport, energy, technology and greenhouse gas emissions. The questionnaire asked respondents, as professionals, to rate willingness to accept these measures, ranging from (1) certainly oppose to (5) certainly accept. Factor analysis of the five measures led us to sort them into two categories: response alternatives with mean values *above* 3.4 are interpreted as 'low cost' (information, subsidies and regulation), while those with mean values of 3.4 and *lower* are treated as 'high cost' (taxes and emissions trading).

7   *Environmental concern* was measured in three ways. First, four environmental problems were listed: vehicle emissions, oil discharges along the coastline, ozone depletion and climate change. Respondents were asked to estimate (on a Likert scale) to what extent they consider each item as ranging from (1) not a big threat to (5) a very big threat. Second, respondents were asked whether they think that 'Climate change will become a serious problem for your organization/for the nation', with answers ranging from (1) not serious at all to (4) very serious. Third, respondents were presented with four statements about climate change as a relevant topic for discussion within their own organization. Respondents were asked to indicate their views on a five-point Likert-like scale ranging from (1) totally agree to (5) totally disagree.

8   Our study included an analysis of respondents' allocation of responsibility to their own organization or to other levels (local, national and global). However, this did not add significantly to the explanation of acceptance of low-cost or high-cost strategies.

9   The face-to-face interviews were conducted during 2004. Based on analyses of the survey, the 19 semi-structured follow-up interviews were held with decision-makers in local governments in four strategically selected municipalities in the Greater Gothenburg region. Our criteria for selecting these four local governments were:

- size of the municipality;
- main sources of GHG emissions; and
- climate *problematique*/capacity of local administration.

The four municipalities – Gothenburg, Stenungsund, Härryda and Öckerö – are all members of the Gothenburg Regional Association of Local Authorities (*Göteborgsregionens Kommunalförbund*, or GR), with 13 members (see GR, 2005a, b). With nearly 500,000 inhabitants, the city of Gothenburg is the dominant local governmental actor in the region. Stenungsund (with a substantial petrochemical industry) and Härryda (with the Gothenburg airport) are suburban municipalities with 22,000 and 30,000 inhabitants, respectively. Öckerö is an archipelago municipality with only 12,000 inhabitants. For a more detailed account of these municipalities, see von Borgstede and Lundqvist (2007).

REFERENCES

Alvesson, M. (2002) *Understanding Organizational Culture*, Sage, London

Bastianoni, S., Pulselli, F. and Tiezzi, E. (2004) 'The problem of assigning responsibility for greenhouse gas emissions', *Ecological Economics*, vol 49, pp253–257

Biel, A. (2003) 'Environmental behaviour: Changing habits in a social context', in Biel, A. and Mårtensson, M. (eds) *Individual and Structural Determinants of Environmental Practice*, Ashgate, Aldershot, UK

Brown, J. S. and Duguid, P. (1991) 'Organizational learning and communities-of-practice: Towards a unified view of working, learning and innovation', *Organization Science*, vol 2, pp40–57

Bruff, G. E. and Wood, G. P. (2000) 'Making sense of sustainable development: Politicians, professionals, and policies in local planning', *Environment and Planning*, vol 18, pp593–607

Cabinet Bill 2001/02:55 (2001) *Sweden's Climate Strategy*, Parliamentary Record, Stockholm, Sweden

Dawes, R. M. and Messick, D. M. (2000) 'Social dilemmas', *International Journal of Psychology*, vol 35, pp111–116

Devos-Comby, L. and Devos, T. (2001) 'Social norms, social value, and judgments of responsibility', *Swiss Journal of Psychology*, vol 60, pp35–46

Diekman, A. and Preisendörfer, P. (2003) 'Green and greenback: The behavioral effects of environmental attitudes in low-cost and high-cost situations', *Rationality and Society*, vol 15, pp441–472

GR (Gothenburg Association of Local Authorities) (2005a) *The Göteborg Region Association of Local Authorities (GR)*, available at www.gr.to/grenglish.shtm

GR (2005b) *Folkmängd 1 jan 2006 Göteborgsregionens kommunalförbunds medlemskommuner*, available at www.gr.to/nyheter/pressmeddelanden/folkmangd_051231.pdf

Haslam, S. A., Powell, C. and Turner, J. C. (2000) 'Social identity, self-categorization, and work motivation: Rethinking the contribution of the group to positive and sustainable organizational outcomes', *Applied Psychology*, vol 49, pp319–339

Hooghe, L. and Marks, G. (2003) 'Unraveling the central state, but how? Types of multi-level governance', *American Political Science Review*, vol 97, pp233–243

Hopper, J. R. and Nielsen, J. M. (1991) 'Recycling as altruistic behavior: Normative and behavioral strategies to expand participation in a community recycling programme', *Environment and Behavior*, vol 23, pp195–220

March, J. G. and Olsen, J. P. (1989) *Rediscovering Institutions: The Organizational Basis of Politics*, Free Press, New York

Messick, D. M. and Brewer, M. B. (1983) 'Solving social dilemmas: A review', in Wheeler, L. and Shaver, P. (eds) *Review of Personality and Social Psychology, vol 4*, Sage, Beverly Hills, CA

Newman, J. and Nutley, S. (2003) 'Transforming the probation service: "What works", organizational change and professional identity', *Policy and Politics*, vol 31, pp547–563

Nilsson, A., von Borgstede, C. and Biel, A. (2004) 'Willingness to accept climate change strategies: The effect of values and norms', *Journal of Environmental Psychology*, vol 24, pp267–277

Nordlund, A. M. and Garvill, J. (2003) 'Effects of values, problem awareness, and personal norm on willingness to reduce personal car use', *Journal of Environmental Psychology*, vol 23, pp339–347

Pillutla, M. M. and Chen, X.-P. (1999) 'Social norms and cooperation in social dilemmas: The effects of context and feedback', *Organizational Behavior and Human Decision Processes*, vol 78, pp81–103

Powell, W. W. and DiMaggio, P. J. (1991) *The New Institutionalism in Organizational Analysis*, University of Chicago Press, Chicago

Selden, S. C., Brewer, G. A. and Brudney, J. L. (1999) 'Reconciling competing values in public administration – Understanding the administrative role concept', *Administration and Society*, vol 31, pp171–204

SFS (Swedish Code of Statutes) 1998:1276 (1998a) *Trafikförordning [Road Traffic Ordinance]*, Swedish Government Office, Stockholm

SFS 1998:808 (1998b) *Miljöbalken [Environmental Code]*, Swedish Government Office, Stockholm

SFS 2004:657 (2004) *Förordning om utsläpp av koldioxid [Ordinance on CO$_2$ Emissions]*, Swedish Government Office, Stockholm

Skagersjö, B. (2002) 'Miljözon för tung trafik i Stockholm, Göteborg, Malmö och Lund', Swedish Association of Local Authorities and Regions (SALAR), Stockholm, www.skl.se/artikel.asp?A=2249&C=1322

SOU (Swedish Government Official Reports) 2000:23 (2000) *Förslag till Svensk Klimatstrategi [Proposals for a Swedish Climate Strategy]*, Ministry of Environment, Stockholm, Sweden

Tait, M. and Campbell, H. (2000) 'The politics of communication between planning officers and politicians: The exercise of power through discourse', *Environment and Planning A*, vol 32, pp489–506

Uzzell, D. L. (2000) 'The psycho-spatial dimension of global environmental problems', *Journal of Environmental Psychology*, vol 20, pp307–318

Van Vugt, M., Van Lange, P. A. M. and Meertens, R. M. (1996) 'Commuting by car or public transportation? A social dilemma analysis of travel mode judgements', *European Journal of Social Psychology*, vol 26, pp373–395

Vining, J. and Ebreo, A. (1992) 'Predicting recycling behavior from global and specific environmental attitudes and changes in recycling opportunities', *Journal of Applied Social Psychology*, vol 22, pp1580–1607

von Borgstede, C. and Lundqvist, L. J. (2006) 'Organizational culture, professional role conceptions and local decision-makers' views on climate policy instruments', *Journal of Environmental Policy and Planning*, vol 8, pp279–292

von Borgstede, C. and Lundqvist, L. J. (2007) 'Whose responsibility? Swedish local decision-makers and the scale of climate change abatement', *Urban Affairs Review* (forthcoming in 2007)

von Borgstede, C. and Zannakis, M. (2005) 'Local governance of climate change: The impact of norms on involvement in collaboration projects', Paper presented to the Seventh NESS Conference, Göteborg University, Sweden, 15–17 June

7

Policy Effectiveness and the Diffusion of Carbon-Free Energy Technology: The Case of Wind Power

Patrik Söderholm, Maria Pettersson, Kristina Ek and Gabriel Michanek

INTRODUCTION

The energy production processes introduced during the 20th century – most notably those relying on the combustion of fossil fuels – have given rise to negative impacts on the global climate. Somewhat paradoxically, policy-makers worldwide now hope that future technological developments will solve the problems that technical change has caused in the past. This requires policy efforts in the energy sector to be heavily focused on innovation and technology diffusion activities as a complement to policies explicitly addressing the reduction of carbon emissions (such as tradable emission rights and fuel taxes) (Jaffe et al, 2005). In this chapter we focus on renewable energy penetration in the electric power sector, particularly the development of wind power. Investments in new carbon-free energy technology face a number of economic, political and institutional hurdles, which, in turn, may motivate the use of public support schemes aimed at speeding up the technology diffusion process (see, for example, Fisher and Newell, 2004). Nevertheless, in order to design efficient policy instruments in the field, a proper understanding of the economic and institutional conditions that govern technology diffusion in the electric power sector is needed.

Our purpose is to use the wind power example to illustrate important challenges to increased diffusion of carbon-free technology in modern society. In doing this, we provide a synthesis of a number of research undertakings and combine a quantitative analysis of innovation and diffusion in the European wind power sectors with an in-depth case study of the experiences of wind power development in Sweden. Most notably, the first part of the chapter permits quantitative tests of: the impact of public policy and cost-related factors on the developments of wind power in Sweden, Denmark, Spain, Germany and the UK; and whether wind power diffusion will differ depending upon the public support scheme used. The second qualitative part of the chapter permits us to gain an understanding of the economic, political and legal conditions that face a wind power investor in Sweden, and also to put the Swedish situation into a broader perspective by comparing the most critical institutional conditions with those existing in Denmark.

Although an extensive theoretical literature on technology diffusion exists, empirical applications based on explicit quantitative tests are rare (Jaffe et al, 2000). Some exceptions include, for instance, Hassett and Metcalf (1995), Jaffe and Stavins (1995) and Koomey et al (1996). Still, these studies deal primarily with the diffusion of end-use energy saving equipment, while the focus here is on the diffusion of an environmentally benign energy supply technology. Past research efforts on the diffusion of wind energy *per se* have mostly been case studies drawing extensively on qualitative evidence in individual countries (e.g. Bergek, 2002 (Sweden); Bird et al, 2005 (US); García-Cebrián, 2002 (Spain); Wolsink, 1996 (The Netherlands)), while quantitative (econometric) studies have relied almost exclusively on so-called learning curve analysis. The latter type of studies investigates to what extent capacity expansions, spurred by research and development (R&D) support and learning by doing, lead to cost reductions. However, in practice, innovations – and, thus, cost reductions – do not automatically lead to increased diffusion of the technology. McVeigh et al (2000) show that even though the costs of renewable energy technologies have fallen far beyond expectations, they have failed to meet expectations with respect to market penetration. These results suggest that:

- the costs of the traditional power sources have fallen as well (e.g. Claeson Colpier and Cornland, 2002); and (equally important)
- apart from cost disadvantages, there exists additional legislative and institutional obstacles to renewable energy diffusion, which so far are only partly understood.

The latter implies that renewable energy policies must address not only financial support but also institutional reforms, legal actions and public acceptance issues.

Our analysis relies on two different types of methodological approaches. The first part draws heavily on the work by Jaffe and Stavins (1995), who developed a rational choice model of technology diffusion that can be applied econometrically. For our purposes, the model specification aims at permitting an analysis of the variations in installed capacity of wind power across five countries (Sweden, Denmark, Germany, Spain and the UK), as well as over time. We also complement this diffusion model with a model of innovation (learning curve analysis). In this way we acknowledge not only that costs matter for diffusion, but also that diffusion is a necessary condition for learning and, ultimately, cost reductions. Thus, diffusion and innovation need to be analysed simultaneously, and policy will affect both. The model is estimated using an unbalanced panel data set covering five Western European countries over the time period of 1986 to 2001. The results from this empirical work indicate, among other things, what the most important factors are affecting the diffusion of wind power and innovation in the power sector, and increase our under-standing of:

- the main driving forces behind the development of wind power in Europe; and
- the extent to which different subsidy systems are more effective than others in promoting diffusion and innovation in this sector (Ek and Söderholm, 2005; Söderholm and Klaassen, forthcoming).

However, it must be stressed that this model approach is also limited in its characterization of the institutional conditions affecting wind power develop-ment.[1] There thus exists a need to complement this broad picture with an in-depth case study of the experiences in a specific country. In our analysis of Swedish wind power, we approach the issue from the perspective of a power generator who considers investing in new wind turbines. This implies that the economics of Swedish wind power is assumed to be affected not only by technology-specific costs and public-support schemes, but also by stakeholder interests, as well as the legal provisions governing the assessment of the environ-mental impacts of wind turbines and the planning procedures for their location. The adopted power-generator eye view of the investment decision process enables us to explicitly analyse the different types of economic, legal and polit-ical uncertainties that face a wind power investor in Sweden, and point to measures that can be implemented to reduce these uncertainties.

WIND POWER IN EUROPE: EXPLORING THE SUCCESS AND FAILURE STORIES

The development of wind power in selected European countries

Figure 7.1 displays the development of wind power capacity in five Western European countries. The choice of countries is motivated, first of all, by the fact that the development of wind power differs among these countries. Germany, Denmark and, more recently, Spain have all experienced consider-able increases in the installed capacity of wind turbines, while the corresponding developments in Sweden and the UK have been more modest. For instance, in 1991, Spain and the UK had more or less the same amount of wind energy

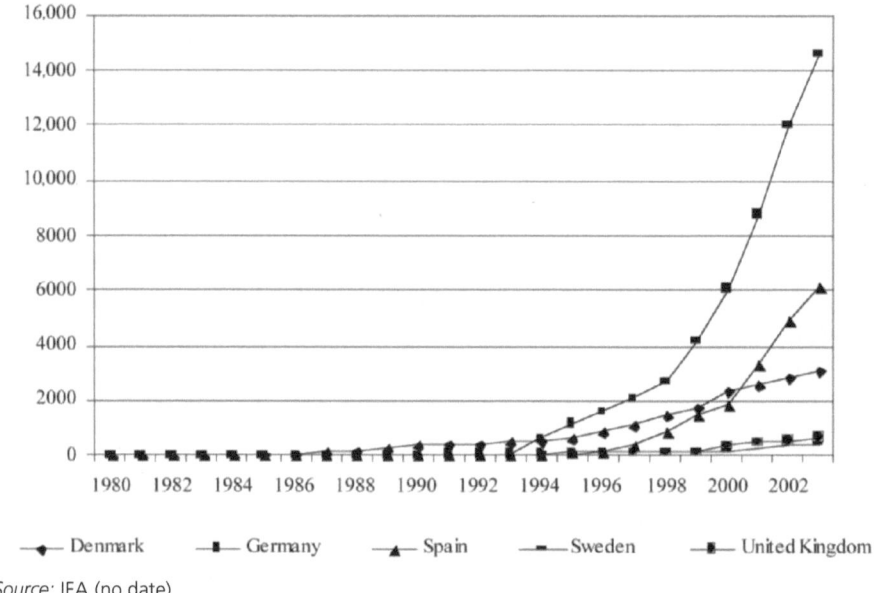

Source: IEA (no date)

Figure 7.1 *Installed wind power capacity in selected European countries (MW)*

capacity installed (around 4MW to 5MW); but in 1999, Spain's wind capacity amounted to 1584MW, while it was only 344MW in the UK. In Sweden, wind power production increased by over 700 per cent over the time period of 1994 to 2002, albeit from a very low level. Thus, in spite of this relative increase, in 2004 the share of domestically generated wind power out of total Swedish electric power supply was only 0.5 per cent. This corresponds to a power generation of about 0.6TWh, far below the Swedish government's policy goal of 10TWh by the year 2015 (Swedish Government, 2002). In Sweden's neighbouring country Denmark, however, wind power's share of total power generation is currently well above 10 per cent.

Moreover, the support systems for wind power differ across the five countries. The UK is the only country that has relied on a competitive bidding system (the so-called Non-Fossil Fuel Obligation). In this system, calls for tenders were made at alternating intervals. Wind power is given a quota. The providers of the lowest asking prices are given contracts, and the contract price received by all wind generators equals the bidding price of the marginal producer. In the fixed feed-in price systems prevailing in Denmark, Germany and Spain (although with some variations), a long-term minimum price is guaranteed *ex ante* for electricity obtained from wind power. In Sweden, different feed-in tariffs (most notably the so-called 'environmental bonus') have also been used to encourage wind power generation. However, the tariff rates were 'renegotiated' annually, giving rise to substantial uncertainties about the long-run economics of Swedish wind power.[2]

Most analysts conclude that the fixed feed-in tariff schemes have had the greatest success in promoting the use of wind electricity since they reduce uncertainty and make it easier for wind energy producers to obtain bank financing (e.g. Meyer, 2003). Still, it is unclear whether differences in wind power diffusion rates between, say, the UK and Denmark are due to the support systems or to other factors such as variations in planning procedures and/or local opposition. Moreover, the impact on innovation activities and, thus, on cost reductions may also differ depending upon the support scheme chosen (see, for example, Menanteau et al, 2003).

Empirical results from the simultaneous innovation–diffusion model

Given this inconclusive situation, we provide quantitative tests of the impact of wind support schemes on technology diffusion and on innovation activities. Figure 7.2 summarizes the estimation results from the quantitative innovation–diffusion model (see Söderholm and Klaassen, forthcoming, as well as Ek and Söderholm, 2005, for details). The results confirm the notion that innovation and cost reductions – spurred partly by public R&D support – are necessary conditions for the successful diffusion of wind power; but the opposite is also true. A wind turbine is not only built because it has become cheap and efficient; it is also true that it becomes cheap *because* it is built and operated (i.e. the learning effect).

Furthermore, the role of price subsidies is important for the diffusion of wind power; but there is a need to carefully design the time development of the support. Increases in the feed-in price for wind power promote diffusion of wind capacity, which, in turn, encourages learning and generates cost reductions. However, there also exists a direct negative effect of feed-in price increases on learning. The reasons for this are that high feed-in prices:

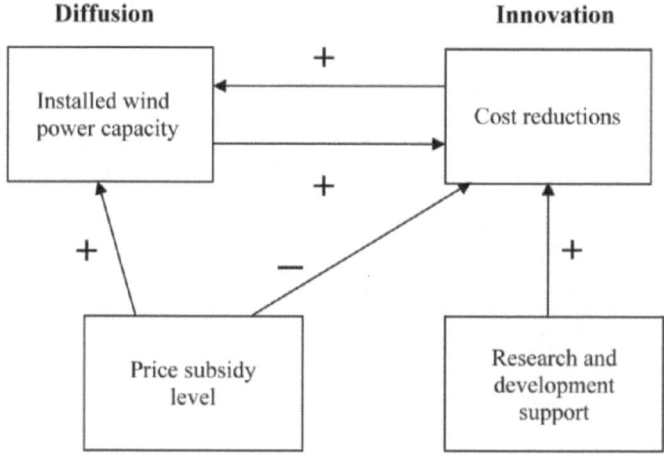

Source: chapter authors

Figure 7.2 *Illustration of wind power innovation–diffusion model results*

- induce wind power producers to select high-cost sites (e.g. locations with expensive grid connections and/or poor wind conditions); and
- discourage the competitive pressure from other energy sources – as a result, innovation activities become less attractive.

This notion has an important policy implication since it suggests that there is an opportunity cost in the promotion of new technologies. Thus, clearly announced gradual decreases in feed-in tariff levels over the lifetime of the wind turbine may be an important element of an efficient renewable energy technology policy. Recent policy developments also move in this direction. The new German Renewable Energy Sources Act of 2000 stipulates decreasing feed-in tariffs over the years in order to take into account technical progress over the lifetime of the turbines. The Danish Council for Sustainable Energy has proposed a similar arrangement for renewable energy sources in Denmark.

Moreover, we find limited support for the notion that the impact on wind turbine investments of a marginal increase in the price subsidy level will differ depending upon the type of support system used.[3] Furthermore, we found no support for the hypothesis that the different price support systems induce varying incentives for cost reduction. These results put in doubt the common assertion that the differing wind power developments are primarily the result of the design of implemented policy instruments. In addition, wind conditions are no worse in Sweden compared to, say, Denmark or Germany. The price subsidy levels do not differ significantly across the countries (e.g. Cerveny and Resch, 1998), and modern wind turbines can be bought on the global market (most notably from Denmark). However, what does appear to differ between the countries is the consistency with which the national wind power policies have been implemented. The economic and policy-related uncertainties that face a wind turbine investor vary heavily across countries in terms of both type and size. The same is true for the public's view regarding wind power development and the legal possibilities to hinder wind turbine installations at the local level (e.g. Reiche and Bechberger, 2004). In sum, a comprehensive analysis of the prospects for future wind power development must not only address the relative costs of wind power generation and the impact of the different policy instruments on these costs. It must also deal with the uncertainties that are created by the regulatory and legal systems, as well as the impact of public perception on wind power development.

GLOBAL POLICIES AND LOCAL OBSTACLES: THE CASE OF WIND POWER IN SWEDEN

In this section we analyse the potential for future wind power development in Sweden, with special emphasis on important institutional obstacles to further diffusion of wind turbines in the country.[4] In line with the above methodological discussion, we assess the economics of wind power investments in close conjunction with an analysis of the legal, attitudinal and policy-related uncertainties that face a wind power investor in Sweden.

The impact of national energy policy on the economics of wind power

In 2002, a national planning goal of a yearly wind power generation of 10TWh by 2015 was adopted. Before this planning goal was introduced, Sweden's wind power policy was characterized by soft formulations stating that wind power should be promoted in the Swedish energy system, without explicitly stating when and how much (Åstrand and Neij, 2006). Although the cost of producing wind power has declined substantially during the last two decades, public support is still generally needed to make investments in wind turbines commercially attractive. In the past, investment and production subsidies dominated the policy portfolio used to encourage wind turbine investments. However, a green certificate system for renewable energy was introduced in 2003. Its aim is (similar to that of competitive bidding systems) to secure a predetermined market share for renewable electric power sources, but also to promote cost-effective competition between the different types of renewable energy sources (Swedish Government, 2003). The new system has replaced previous investment subsidy programmes and will gradually replace the production subsidies, which will be lowered annually and (in the case of onshore wind power) be completely abandoned in 2010. It should also be noted that in the past no carbon tax was paid for fuels used in the Swedish power sector. However, with the 2005 introduction of European Union (EU) trade in emissions allowances for carbon dioxide, power-related carbon emissions also carry a price.

Table 7.1 summarizes the levelled (lifetime) cost estimates for different new power generation technologies in Sweden (for commissioning in 2003) as reported by the Swedish electricity research institute *Elforsk*, including one onshore and one offshore wind power alternative. The costs for wind power include all investment costs (turbine, electrical installations, foundation, etc.), but ignore the highly site-specific costs related to connections to the electric grid.[5] Overall, the cost figures show that in the absence of taxes and subsidies, gas-, coal- and some hydro-based power are the cheapest alternatives (although further development of new large-scale hydropower is strongly restricted according to Swedish law). However, when existing taxes and subsidies are added and subtracted from the private costs, the competitive positions generally change in favour of wind power. Specifically, the far right column in Table 7.1 shows the different levelled costs after an electricity certificate price of 0.15 Swedish kronor (€0.016, or £0.011) per kilowatt hour and the discounted value of the future time-declining environmental bonus have been subtracted from the wind power costs; and after the taxes charged on sulphur and nitrogen emissions have been added to the fossil-fuelled power generation alternatives.[6] As a result of the policies implemented, wind power appears to be one of the most attractive new power generation investments in Sweden.

Nevertheless, the engineering cost figures presented in Table 7.1 build on specific assumptions about discount rates and subsidy levels. They also, therefore, neglect the role of different uncertainties related to the policies and institutional frameworks that govern wind power development. For this reason,

Table 7.1 *Lifetime generation costs for new power plants in Sweden*

Plant type (CHP = combined heat and power)	Capacity (MW)	Levelled cost (Swedish kronor per kWh)*	
		Without *taxes and subsidies*	With *taxes and subsidies*
Coal – power plant	400	0.39	0.43
Coal – CHP	100	0.30	0.79
Gas – power plant	400	0.30	0.31
Gas – CHP	150	0.31	0.46
Bio-fuel – CHP	80	0.40	0.24
Wind power – onshore	20	0.38	0.20
Wind power – offshore	90	0.41	0.23
Hydropower – low	**	0.23	0.23
Hydropower – high	**	0.36	0.36

Notes: * The levelled cost estimates are based on the use of a 6 per cent discount rate and an economic lifetime of 20 years (except for hydropower, for which the economic lifetime is assumed to be 40 years). The costs for producing hydropower tend to vary significantly depending upon location, and for this reason two estimates are presented, where the actual cost is assumed to lie somewhere between these two extremes. Since the Swedish government has decided to gradually phase out nuclear power and, thus, no *new* nuclear plants are planned, this option is not included here.
** It is not possible to provide a meaningful value for hydropower capacity since much depends on the site used.
Source: Bärring et al (2003)

Söderholm et al (2007) analysed the impacts of tradable emission rights for carbon dioxide and the green certificate system under different rate-of-return requirements on the relative cost structure of wind power. The levelled cost of gas-fired power generation served as a benchmark in this analysis. There are two major conclusions from these simulations:

1   The allowance system alone would provide a sufficiently strong policy instrument to put wind power on an equal footing with gas-fired power *only* if the carbon trade is expanded to additional sectors than the ones currently involved (e.g. heat and power, iron and steel, pulp and paper, etc.), *and/or* a stricter cap on total emissions is introduced.
2   Overall, the green certificate system has (so far) provided a strong economic stimulus to wind power. The average certificate price over the time period of September 2003 to September 2004 was 0.22 Swedish kronor (€0.024, or £0.016) per kilowatt hour (Swedish Energy Agency, 2004), and at such prices wind power will be cost competitive compared to gas-fired power even at relatively high discount rates.

Nevertheless, the certificate system has been connected with a number of uncertainties – such as price fluctuations – implying that the risk-adjusted discount rate has been high. Most importantly, perhaps, while the economic time horizon of a wind power project is generally (at least) 20 years, the green certificate system has had a much more limited time horizon; it was planned to exist at least until the year 2010; but after that it was unclear what would follow. This signalled a lack of political commitment and increased the economic risks faced by investors; but in 2006 a proposal for an extended system was put forward by the Swedish government (Swedish Government, 2006).

An important result of the analysis is also that wind power loses competitive ground from the use of higher rate-of-return requirements. This is because the capital costs involved in wind power development form a sizeable part of the total levelled costs, and the higher the uncertainties about the future rate of return of the investment are, the less competitive wind power will be. For instance, both bio- and gas-fuelled power are less capital intensive (in relative terms) and will thus benefit from increased uncertainties about market and policy developments, as well as about the outcome of planning and permitting procedures. In addition, the presence of an unstable policy environment tends to favour investments in – and intensified use of – *existing* capacity at existing sites. A number of renewable power alternatives that involve investments – and resulting production increases – in existing capacity are eligible for certificates. These include, most notably, the substitution of biomass for coal in existing combined heat and power (CHP) plants and the upgrading of existing hydropower. This introduces a large degree of path dependence in the energy system, and harms all new investments in power generation technologies in Sweden. The main advantage of these options lies in the fact that the investment costs of the existing plants are sunk, and they will compete with new capacity largely on the basis of their variable costs. The greater the difference between the total cost of a new plant and the variable cost of an existing plant, the greater the incentive for better and more intense use of the existing plant. This is typically the case when new wind power competes with existing hydropower or nuclear energy.[7] In sum, our analysis suggests that it is generally more efficient to promote wind power by reducing the uncertainties about future regulations and policies than providing *additional* economic incentives by introducing new policy instruments or strengthening existing ones.

General public support for the National Wind Energy Policy

The occurrence of local resistance towards planned wind farms is often referred to as an important obstacle to increased wind power capacity in Sweden and elsewhere. Fears of visual intrusion, noise and land devaluation often explain these negative opinions. However, in spite of the existence of local opposition, the experiences in Sweden (and in many other countries) are that lay people generally express a positive attitude towards wind power (e.g. Krohn and Damborg, 1999; Ek, 2005). For this reason, the occurrence of local resistance towards wind power development is often explained by the so-called not in my backyard (NIMBY) syndrome.[8] This explanation has, however, been criticized for being too simplistic (e.g. Wolsink, 2000). Local resistance may, instead, often express suspicion towards the people or the company who want to install the turbines or a rejection of the process underlying the decision to build new plants, rather than a rejection of the turbines themselves. Results from interviews with people living close to wind power installations in the south of Sweden also emphasize the role of collaborative approaches and the benefits of involving the local population in the early stages of the planning of wind turbines (e.g. Hammarlund, 1997; Swedish Energy Agency, 1998).

This illustrates the importance of analysing public attitudes towards wind power in close conjunction with the legal and institutional frameworks that affect the development of wind power. Legal rules for wind power siting (as well as resulting court decisions) generally aim at finding a proper balance between different interests in society. Such rules will therefore largely determine the extent to which any negative opposition will influence wind power siting decisions. Here, we add to the empirical evidence on the public's attitudes towards wind power in Sweden on the basis of a postal survey carried out in 2002 (for details, see Ek, 2002). The main objective of the survey was to analyse the attitudes towards wind energy in general, as well as the perception of the different attributes of wind power.

When asked to state their general attitude towards wind power, only 10 per cent of the respondents expressed a negative stance, while nearly two-thirds (64 per cent) were positive. The likelihood of finding an individual who is positive towards wind power differs with attitudinal and socio-economic characteristics (see Ek, 2002, for details). Support for wind power tends to decrease with age and income, while education level does not have any statistically significant impact. The negative correlation between income and the probability of a positive attitude is somewhat unexpected, and contradicts results from other studies (e.g. Roe et al, 2001; Zarnikau, 2003). One possible explanation might be that individuals with higher income put less significance on the positive employment effects associated with wind power installations. No support is found for the hypothesis that differences in attitudes vary with respect to own experiences of wind power installations. Individuals with wind turbines in sight of their home or summer house did not appear to perceive wind power in a significantly different way compared to individuals without such experience. These results thus lend no support to the NIMBY hypothesis. Nevertheless, it should be clear that this is only a 'weak' test of this hypothesis; the NIMBY phenomenon (to the extent that it exists) is likely to be particularly prevalent prior to the construction of a new wind turbine. Furthermore, wind power is perceived as an environmentally benign electric power source, and people who act to protect the environment (here, those stating that they regularly buy 'green' products) are also more likely to express support for wind power.

We also analysed how respondents view social choice in the energy and environment field, and if these views affect their attitudes towards wind power. Two issues relating to social choice were examined. The *first* issue dealt with the respondents' willingness to accept trade-offs between environmental quality, on the one hand, and economic benefits, on the other. The results suggest that people who reject the idea of such a trade-off are more likely to express a positive attitude towards wind power than those who wish to strike a balance between economic and environmental goals, and thus are more willing to give up environmental benefits for, say, lower electricity prices. The *second* social choice issue dealt with the respondents' view on private versus public choices. A distinction here was made between those who expressed support for the idea that the political system should form the basis for decisions on the introduction of 'green' electricity, and those who believed that the market mechanism should determine the extent to which 'green' electricity was introduced in Sweden. The results indicated that the more people stressed the importance of the political system, the more likely they were to express a positive attitude towards wind power. This effect was, however, not statistically significant.

Energy policy documents typically stress the environmental advantages of wind power compared to other power sources, particularly the fact that it does not generate emissions of any harmful substances. Nevertheless, much of the opposition towards wind power has targeted different negative attributes of wind power, such as visual intrusion, noise pollution and impacts on flora and fauna. For this reason, we also comment briefly on the results from a so-called choice experiment whose aim was to elicit the respondents' preferences towards the different *attributes* of wind power.[9] When the attributes included in the experiment were selected, the results from previous research efforts constituted an important input (e.g. Hammarlund, 1997; SOU, 1999; Pedersen and Persson Wayne, 2002). According to this research, the amenity effects are of major importance for the public's perception of wind turbines. The attributes included in the experiment to capture the attitudes towards the visual impacts were the location (onshore near the coast, onshore in the mountains and offshore), and the height and grouping of turbines (large wind farms, smaller groups and separately located turbines). A noise attribute and a cost attribute (changes in the electricity price) were also included in the choice scenario. Respondents were asked to choose between two alternatives of wind power, each associated with different environmental attributes and prices. In order to make the choice task easier, the different levels of the included attributes were briefly described and illustrated in combination with some reference levels.

Our results indicate that Swedish electricity consumers are highly cost conscious. Furthermore, the findings confirm previous research results stating that the visual impacts are of vital importance. The location attribute of wind turbines appears to have the largest impact on the utility of the respondents. Our results suggest that wind power located offshore is considered an environmental improvement, while a location in the mountains is considered an environmental deterioration (compared to a location onshore near the coast). In addition, small wind farms are considered a change for the better, while large farms are considered a change for the worse (compared to separately located wind turbines). Finally, our results imply that reduced noise levels would increase the utility of the average respondent; but this impact was not statistically significant.

In sum, the Swedish public generally expresses a positive attitude towards wind power. This positive attitude tends to be correlated with a willingness to defend – and act on – environmental values. This suggests that there appears to be a relatively strong support for the Swedish energy policy objective to support increased diffusion of wind power. Nevertheless, this support is not always visualized at the implementation stage. The results from our choice experiment indicate a number of strategies that can be used to reduce any negative perceptions of wind power. To minimize environmental disturbances, new schemes should primarily be located offshore, while large wind farms onshore or in the mountains should be avoided. Thus, even though offshore wind power is generally more expensive than land-based turbines, this is, at least partly, offset by the lower risk of public opposition for offshore installations.

Legal preconditions for implementing the National Wind Energy Policy at the local level

The installation of wind turbines is largely conditional upon the requirements of the law. Swedish law grants a significant amount of discretion to the local authorities. In the case of wind power development, the system of rules governing the use of land (and water) areas, as well as the assessment of the environmental impacts of turbines, is of particular interest. In Sweden, the development of wind turbines is primarily regulated in the Environmental Code. The code states that the installation of large and medium-sized turbines can be permitted only if they are in compliance with certain environmental requirements, among which are the basic and special resource management provisions and the so-called localization requirement. Also of importance are the rules on physical planning in the Planning and Building Act and the specific legal prerequisites for *offshore* wind turbine installations. The following analysis shows that although national energy policy promotes increased reliance on wind power, the implementation of the above legislation at the local level tends to take only limited account of these national policy aspirations.

One should note, of course, that the regulations on environmental assessment and territorial planning are designed to promote an *overall* efficient use of resources and to secure certain legal rights. Their purpose is *not* to support the rapid diffusion of wind power. Nevertheless, it is useful to analyse these legal provisions from a wind power investor eye view, not least since this approach permits us to highlight some of the (explicit and implicit) trade-offs made in the intersection between national energy policy goals and local priorities.

Regulations concerning the use of land and water areas

The basic resource management provisions in the Environmental Code include general provisions for assessing different land-use interests. However, they also provide legal 'protection' for areas related to certain interests of particular importance to the public interest (e.g. areas particularly suitable for energy production, nature conservation and/or recreational activities).

The weighting provisions are formulated quite broadly. They thus provide much room for different interpretations regarding the legal application, as well as the actual content, of the provisions. For instance, the main rule that 'priority shall be given to use that promotes proper management from the public interest point of view' merely implies that:

- a long-term perspective should be applied on all land- and water-use issues; and
- the interests of the public should take precedence over any private interests, although a combined use should always be considered.

'Wherever' possible, the legally 'protected' areas should be protected against activities that may *significantly affect or damage* the character of an area or are *prejudicial* to its use. The protection is, however, relatively weak. Since areas may be 'suitable' for more than one purpose, the basis for the assessment is the very vague general rule quoted above. At best, areas may be designated as 'national interests' for wind power production, implying that the areas *shall* be protected against (in this case) *prejudicial* (i.e. constraining) activities.

Our analyses of case law confirm that the prerequisites for wind power development provided by the basic resource management provisions are unpredictable regarding the possibility of averting obstructive activities and of (explicitly) promoting wind power (Söderholm et al, 2007). For energy policy purposes, this implies that there exists a need to strengthen the weight given by designating areas as of national interest. Whether an area is, in fact, of national interest for the designated purpose is ultimately, however, a matter for the court to decide; government agencies cannot make legally binding decisions on this issue. All considered, it is difficult to foresee to what extent wind turbines will be granted permission, and this vagueness adds to the uncertainties faced by an investor.

The Swedish Environmental Code also outlines *special* resource management provisions. These protect geographically delineated areas from exploitation and environmental interferences due to their natural and cultural values. Such an area is, *in its entirety*, of national interest, which implies that the weighting has already been made and that, in a competitive situation, precedence should be given to the protected interests. Wind turbines can only be developed in these types of areas if they *meet no hindrance* by the area provisions and do not *significantly damage* the protected values. In the assessment of the latter impacts, the *total* natural and cultural values are in focus. Thus, even if parts of the protected areas were to be significantly damaged by a specific activity, the rule may not prevent this activity unless the total values of the area are affected.

There are, however, some exceptions to this general prohibition. They generally apply to the development of existing urban areas and the local industry. Wind turbines may be of interest in such cases if providing electricity to new residential areas, retail trade and smaller industries, and/or by providing employment opportunities in the establishment and operation phases. Still, our case law analyses show that, in these cases, the management provisions leave the authorities with significant discretionary power.

As noted, wind turbine development is also subject to the 'localization' rule, under which requirements regarding the selection of sites can be brought upon operators. The localization requirement has provided an important obstacle to wind turbine development in Sweden on several occasions. Two issues are of particular concern. First, for permanent (in contrast to temporary) activities, the selected site must be suitable with regard to the objectives of the code and resource management provisions. Second, for *all* activities, sites must be selected so that the purpose of the activity is achieved with 'a minimum of damage or detriment to the environment'. In controversial cases, the latter requirement obliges the operator to undertake an objective assessment of alternative sites. This may, in some cases, imply a very stringent – and even inefficient – obstacle towards installation. The owner of the turbine may not have access to any other site than the chosen one; but if another site is found to better achieve the purpose of the activity from an environmental point of view, a permit cannot be issued unless the costs for altering the location are found unreasonable.

Wind projects in water areas are subject to additional Environmental Code provisions: 'Water operations may only be undertaken if the benefits from the point of view of public and private interests are greater than the costs and damages associated with them.' This social cost-benefit rule has been applied in favour of offshore wind projects, and case law analysis shows also that the Environmental Court of Appeal regards the state subsidies granted to wind power as benefits from the public's perspective in the weighting process. The subsidies are said to reflect the implicit value of attaining an increased share of renewable energy. The government has also explicitly expressed support for this legal interpretation. This could prove to be important for the future of offshore wind power in Sweden since it illustrates how the wind energy interest – as a means to achieving national policy objectives – can be visualized at the implementation stage and weighted against local impacts. Similar legal approaches are, however, lacking in the case of onshore wind power.

Territorial planning regulations and the consequences of the municipal planning monopoly

Even if a wind power project passes the legal hurdles outlined in the Environmental Code, the project must also be in compliance with the physical planning provisions laid down in the Planning and Building Act. The Swedish physical planning system has a significant influence on the potential for a broad implementation of wind power, not least since it, in principle, implies that the municipalities must in some way assent to (i.e. plan for) the establishment of wind turbines at a certain location in order for the installation to actually take place.

The planning process involves balancing between different interests, and it is mainly a matter for the municipal authorities. The balancing principles are, however, vague and leave substantial room for discretion on the part of the local municipalities. Even though the government (represented locally by *länsstyrelserna*, the county administrative boards) is obliged to reject municipal plans that do not take national interests into account, it is ultimately the municipalities which decide whether or not to accept the boards' advice. In practice, the courts appear to pay a lot of attention to the municipal positions in the permitting process, especially if there is intense competition for land areas. All in all, the municipal planning monopoly often leaves substantial room for discretion and for *de facto* ignoring national (and, indeed, international) energy policy objectives.

Previous governmental investigations have (empirically) illustrated the key role played by local governments in the development of onshore wind power. Given an ambiguous national policy towards wind power, this local role has become even more accentuated. Already in 1998, a commission noted that the attitudes of local governments towards wind power development have differed markedly, and that this has often determined the outcome of the permitting procedure (SOU, 1998; Swedish Energy Agency, 2003). A recent more systematic empirical analysis of how planning strategies employed in three Swedish municipalities influence wind power developments looked at how these strategies affected:

- the siting of wind turbines;
- the ownership pattern of the turbines; and
- the role of citizen participation in each municipality.

The three municipalities under study – Laholm, Halmstad and Falkenberg – are of similar size, and possess roughly similar wind conditions and landscape characteristics; but in terms of wind power development the outcomes differ. Most notably, in Laholm, 45 wind turbines (totalling 22MW) are installed, while only 5 turbines (2MW) have been developed in Halmstad (Khan, 2003).

Khan (2003) proposes that the extent to which, and the way in which, territorial planning requirements have been implemented largely explains the varying outcomes. His results suggest that in municipalities where there exists a political will to promote wind power and, thus, to integrate efficiently the diffusion of wind turbines into the planning process, the planning requirements have typically been flexible and simple. Important drawbacks of this approach, however, are that it may not promote an efficient siting of wind turbines and tends to limit the role of citizen and stakeholder participation. While successful in the short run, such a planning approach may, in a longer perspective, create suspicion towards wind power development. In municipalities where politicians and officials are more reluctant to actively promote wind power, the planning requirements have been stricter and citizen participation more extensive. As a result, the installed capacity of wind turbines is low.

For a potential wind power investor, all of this implies considerable uncertainties about the investment conditions. Even though the economic support is the same across the country, the legal obstacles may differ considerably between regions. Of course, the legal framework is designed to address local circumstances and, as such, it serves a good purpose. However, Khan's (2003) study shows that observed differences between the various municipal planning requirements can, to a large extent, be explained by differences in the attitudes of politicians and even local officials (see also Bengtsson and Corvellec, 2005). This implies that Swedish wind power often faces significant local obstacles to its implementation, and that the contribution to the fulfilment of national policy goals will not be paid enough attention.

Finally, the municipal planning monopoly also tends to undermine one of the goals of the recently introduced green certificates system designed to promote a *cost-effective* introduction of renewable electric power. However, this goal will not be accomplished when the local policies regarding wind power projects differ on grounds that cannot be attributed to important environmental and/or economic conditions. In other words, the Swedish green certificates system (at best) ensures necessary, but insufficient, conditions for a cost-effective deployment of renewable electricity sources.

The complex role of stakeholder participation: Legal situation and implications

The interests of those who object to wind turbine installations often gain strong legal protection and thus make the role of stakeholder participation crucial. Swedish law provides for – and encourages – stakeholders to participate in the decision-making process. Applications are sent out for comments, and consultations and public meetings are arranged in connection with the environmental impact assessment, as well as the planning and permitting procedures. The Environmental Protection Agency, the National Board of Housing, Building and Planning, the County Administrative Board, one or several municipalities, individuals and organizations are all invited to participate. In principle, the courts are obliged to independently assess the individual circumstances in each case. They carry out their own evaluation and balancing of interests and draw conclusions consistent with the preconditions in the law. As a consequence, the courts should ignore opinions put forward by a state agency or a municipality if these are in conflict with the court's own judgement. Nevertheless, standpoints taken by state authorities and municipalities regularly influence the final decision. When arguing for a certain conclusion, the court may even refer explicitly to these standpoints. Since the only formally binding legal source – the legal text – does not in any precise way outline how to value and balance the interests involved, it is inevitable that stakeholders' attitudes often gain significant weight in the courts' decisions.

Stakeholder participation is time consuming and may therefore significantly delay the implementation of wind power projects. This problem is accentuated by the appeal possibilities and by the overlap of the permit and planning systems. Assume, for instance, that a large wind turbine is projected offshore close to a city. The Planning and Building Act would normally require a detailed plan and a separate building permit. According to the Environmental Code, two permits are needed, one since the turbine constitutes a (potential) 'environmentally hazardous activity', and another since it represents a 'water activity'.[10] Each of the permitting processes provide for stakeholder participation and for appeals in two additional instances. Although the different permitting procedures are sometimes coordinated, it is obvious that the overall planning and permitting process may take several years, in some cases eight to ten years, particularly in those cases where strong negative attitudes are expressed among relevant stakeholders.

The issue of participation and access to justice is a delicate one. One can argue that the likelihood of obtaining the environmentally 'right' decision will increase with frequent stakeholder involvement. Clearly, it is also a matter of democracy and the necessity to increase the long-term legitimacy for wind power. The previous strongly negative stance among many Swedes towards the nuclear industry was partly due to the permitting arrangements. Safety in nuclear plants was judged according to a special legislation with a procedure that did not permit substantial participation by the public.[11] Still, the issue is quite complex. On the one hand, the Swedish public tends to provide a lot of general support for the national energy goals concerning wind power. On the other hand, other goals can easily be – and, indeed, often are – put in the foreground at the local level. It may thus be equally appropriate to argue that the prospects of a legal process stretching over eight years or more, and the associated costs, will deter too many from even contemplating a wind power project. It thus seems necessary to review the planning and permitting process in order to make it less time consuming *without* compromising essential possibilities for stakeholder participation and access to justice. One alternative could be to increase the presence of local ownership in wind projects. In contrast to the situation in some other European countries (e.g. Denmark and Spain), systematic use of means to promote local participation in wind power projects is generally lacking in Sweden. Before 1991, it was not even possible for small private Swedish investors to get public economic support for turbine installations.

For the prospective wind power investor, another strategy to avoid land-use conflicts and related public criticisms would simply be to install the turbines out of view, preferably offshore. We have seen that offshore installations tend to gain more public support than onshore ones, primarily since the aesthetic and noise-related intrusion is often perceived as less severe. The conflict of interest between national energy policy priorities and local implementation also appears to be less intense in the offshore case. In addition to avoiding the removal of land from other competing uses, offshore installations offer some operational advantages. The wind conditions are generally better; but so far this is offset by the higher capital costs associated with offshore investment. Nevertheless, new wind power projects – in Sweden and abroad – are increasingly planned offshore (e.g. Pasqualetti, 2004).

PLANNING FOR WIND POWER IN DENMARK: A COMPARATIVE PERSPECTIVE

Danish energy policy has for long included different measures to promote the implementation of wind power. At the beginning of the 21st century, as much as 15 per cent of total electric power generation in Denmark was generated with the help of wind power (IEA, no date). This is by far the highest wind power share in the world, and there are many factors which help to explain this successful development (see, for instance, Buen, 2003, for an overview). Here, we focus solely on the Danish wind power planning and permitting processes –

that is, how national energy policy goals are implemented at the local level – and how they differ from the Swedish case.

The establishment of new wind turbines in Denmark is almost without exception regulated within the legal framework of physical planning. The Danish planning system has a hierarchical structure involving three authoritative levels (national, regional and municipal) and four different types of physical plans (national, regional, municipal and local). The overall competence structure implies that the national planning authorities deal with overarching planning issues, as well as the implementation of national planning objectives, whereas the regional and municipal planning authorities handle the planning of the open land and the town areas, respectively.

The function of this hierarchical system is built upon two closely related characteristics: *'rammestyring'* and 'strive for' provisions, which are features of central importance in terms of the prerequisites for implementing national planning objectives, such as increased wind power. *Rammestyring* (framework steering) implies a framework of rules to guide individual decisions. Each level of planning provides the framework within which the lower-level planning may be conducted. For instance, the regional planning authorities must respect the framework set by national planning, and municipal plans must comply with regional planning guidelines. Overarching planning objectives may thus be implemented through the national-level plans and all the way down to the legally binding local-level plans. In other words, the different plans are *vertically integrated* (Basse, 2001), and – as a main rule – regional planning guidelines may not be contradicted by municipal or local plans. Areas designated for wind turbine installations in the regional plan will be appointed for the same purpose in the municipal plan. The *rammestyring* is connected to the 'strive for' provisions, which obliges the planning authorities to *strive* to implement the plans or planning guidelines that they have adopted when exercising authority in accordance with the Planning Act (Tegner Anker, 2001).

A national wind power planning directive was issued in 1999 to secure implementation of the national energy policy objective of reducing carbon dioxide emissions through increased use of renewable energy resources.[12] This directive is to be implemented by means of regional and municipal planning. It stipulates that areas suitable for wind turbine establishments in terms of environmental impacts and energy efficiency should be designated and included in the regional planning guidelines. Municipal and local plans for wind turbine installations may only be established for areas already designated for this purpose in the regional planning guidelines. The regional planning authorities thus have the primary responsibility for wind power planning in Denmark, including the drafting of environmental impact assessment reports. Although areas suitable for wind power purposes should be appointed in the regional plan, the directive does not *oblige* the regional planning authorities to designate areas for installations. However, to ensure that areas suitable for large wind turbines are protected from constructions or installations that may interfere with a *later* establishment of large turbines, comprehensive planning to reserve such areas may be required.

The different frameworks of rules governing the wind power planning process are seemingly comparable in Sweden and Denmark. Both systems are decentralized in terms of far-reaching distribution of competence among several planning levels. Nevertheless, there are some crucial differences in the implementation process. The Swedish system is less precise regarding the content of the rules, as well as the legal application, while the Danish system creates a better potential for implementing national goals. The Danish *rammestyring*, together with the possibility of adopting partly mandatory planning directives, implies that national-level policy objectives may not be overlooked either in the planning process or in the implementation of an adopted plan. The possibilities for effective legal implementation of national energy policy objectives thus differ considerably between Sweden and Denmark. In Sweden, there seems to be a substantial 'gap' between national policy objectives and local implementation decisions, whereas the vertically integrated Danish planning system, in part, prevents the existence of such a 'gap'.

Moreover, the precise regulations and specified prerequisites in the Danish laws and bylaws leave the administrative authorities with less room for discretion than is the case in Sweden. This is not the least due to the fact that the implementation of the Swedish wind energy goal must pass not only the various legal hurdles in the Environmental Code, but also in the Planning and Building Act. Overall, the Danish planning system provides a legal framework that is binding in essential parts. Within this framework, the planning authorities may decide *how*, but only to a more limited extent *if*, wind turbines are to be installed.

Just as in the Swedish case, the general attitude towards renewable energy in Denmark is positive. About 80 per cent of the Danes support the idea that promotion of renewable energy sources should be given a higher priority in Danish energy policy. As many as 82 per cent of the Danes are in favour of increased reliance on wind power (Danish Wind Industry Association, 1993). Nevertheless, any local opposition should also be addressed in the decision-making process, and one of the main objectives of the Planning Act is to encourage citizen participation (Tegner Anker, 2001). However, in contrast to the Swedish case, a maximum of two permits are needed in Denmark (a local plan and, depending upon the location of the turbine, in some cases a so-called zone permit). Our analysis of Danish case law also suggests that in order to voice any negative attitudes towards planned wind projects, it is important to get involved early in the decision-making process, while it is easier in Sweden to prevent the installation of turbines at later stages. Clearly, this means that the economic risks facing the wind turbine investor will be more pronounced in the Swedish case (Pettersson, 2006).

Moreover, Denmark has taken active measures to increase the level of local stakeholders and ownership in the wind projects. Since 1979, small private investors can get public economic support for wind turbine investments. Investigations show that people who own shares in wind turbines are more likely to be positive towards wind power compared to people who do not own such shares. Comparing the impact of ownership on wind power installation in Denmark and the UK, Toke (2002) concludes that the presence of Danish co-operatives has been an effective way of enforcing an ambitious central wind-promoting policy, while at the same time avoiding local opposition.

Evidence thus suggests that the different processes of wind power planning in Denmark and Sweden, provide an important explanation for the wide differences in wind power development between the two countries. The strength and design of the public support systems provide only partial explanations; it is just as important to understand the way in which the incentives created by these support systems are 'filtered down' from the national level to implementation at the local level.

CONCLUSIONS AND IMPLICATIONS

The analysis indicates that the strength and design of the different public support schemes for wind power can only to a very limited extent explain important inter-country differences in national wind power developments. Our in-depth study of the potential for future wind power development in Sweden shows, instead, that the existing and planned policy instruments to promote wind power are generally strong enough to make wind power projects economically attractive. However, the economics of wind power is strongly affected by investment uncertainties related to:

- a lack of policy stability;
- public criticism at the local level; and
- the legal provisions governing the assessment of the environmental impacts of wind turbines and the planning procedures for turbine locations.

At the same time, since national and global energy policies, as well as the general public, point out wind power as particularly environmentally friendly, most of the objections to its expansion at the local level also tend to have environmental (or at least aesthetic) origins. The interests of those who object to wind turbine installations gain strong legal protection, and the municipal planning monopoly in Sweden makes it hard for national energy policy goals to come through at the local implementation stage. In Denmark, on the other hand, the planning system provides a legal framework which is binding in essential parts, and within which it is harder for local levels to override national energy policy goals. Compared to its competitors, wind power is one of the power generation technologies that tends to have the most to lose from the risk and uncertainties created by planning regulations that leave much discretion to local authorities. From our analysis one cannot, however, draw the conclusion that Sweden's permitting and planning system should be revised to facilitate increased wind power generation.[13] What it does illustrate, however, is that climate and technology policy may be harder to implement in countries with extensive decentralization of political power. A strengthening of central state authority would possibly facilitate the development of wind power in Sweden.

This also suggests that although public support of wind power may be desirable, the introduction of new policy instruments or the modification of existing ones in countries such as Sweden should be preceded by an evaluation of the institutional framework governing wind power development. The same policy instrument – in terms of both size and design – can induce significantly different developments depending upon the legal preconditions for the siting and assessment of turbines. The success and failure stories of technology support policies can thus not be easily transferred across country borders. This goes for other new technologies, as well; the institutional frameworks governing their gradual diffusion may be fundamentally different when compared across countries.

In addition, climate policy is global in scope, and in Europe there also exist long-term political aspirations to integrate the different types of national support systems for renewable energy (e.g. Midttun and Koefoed, 2003). The presence of significant differences in planning procedures for renewable energy projects may, however, create tensions since stringent conditions in one country will increase the joint aggregate cost of attaining, for instance, the EU target for the share of renewable energy sources. The benefits of climate policy – including wind power diffusion – are largely global in scope; but the costs of implementation are often borne at the local level. This fact may act as an impediment to increased international integration; but it may equally well put pressure on countries to reform local planning and permitting strategies. The analysis of the prospects for each of these two outcomes ought to provide an important area for future research.

The wind power example illustrates an additional general obstacle to increased diffusion of new – and environmentally benign – energy technologies. This obstacle relates to the simple fact that these technologies are new and require greenfield investments. The institutional framework governing energy technology diffusion sometimes suffers from what may be referred to as 'grandfathering', implying that new investments typically face stronger legal and attitudinal obstacles compared to existing production facilities (which may often benefit from significant state subsidies). This introduces a large degree of path dependence in the energy system, and the incentives to increase and extend the use of existing energy production capacity are often understated. Renewable power sources – in particular, wind power – tend to suffer a lot from this policy environment given their high relative capital costs. Over the longer run, greenhouse gas reductions require a fundamental restructuring of the energy system; such a policy endeavour needs to address not only the conditions for new technologies, but also the corresponding conditions for existing technologies. The more profound the uncertainties about future policy developments are, the stronger the incentives to maintain the existing energy system will be.

NOTES

1   The models used take important (time-invariant) inter-country differences in, for instance, permitting procedures and wind conditions into account, but only as a way of avoiding bias in the estimation results (i.e. so-called fixed effects). In other words, little knowledge about the impact of institutional constraints is gained in the analysis.

2   In 2002, the Non-Fossil Fuel Obligation in the UK was replaced by the so-called Renewable Obligation System, which is a tradable renewable certificate system similar to the one introduced in Sweden in 2003. Since the empirical analyses presented here concern the time period prior to 2002, these policy developments are not reviewed in any detail here.

3   Specifically, in Söderholm and Klaassen (forthcoming) it is found that a 1 per cent increase in the price subsidy level implies a 3.9 per cent increase in the installed capacity of wind power in countries employing long-run fixed feed-in tariffs, while the corresponding impact in countries with competitive bidding systems is 1.5 per cent. However, this difference in impacts is only statistically significant at the 12 per cent level.

4   This section draws heavily on an earlier paper written within the COPE (Communication, Organisation, Policy Instruments, Efficiency) programme for research into ways of achieving the Swedish objective of reduced climate impact. See Söderholm et al (2007).

5   This is an important limitation of the analysis. In Sweden, wind power is sometimes constrained by the fact that local grids need to be reinforced before they can deploy new wind power.

6   The taxes and subsidies added in Table 7.1 do not, however, include the impact of the tradable permit scheme for carbon dioxide.

7   The interest in upgrading existing hydropower in Sweden is currently very high among Swedish power producers with the introduction of the green certificate system. A similar development is taking place in the Swedish nuclear industry, in which 600MW of capacity has been added incrementally by improving the use of existing plants (Rogner and Langlois, 2000).

8   The NIMBY syndrome, as a concept, describes people who may well accept the policy that wind turbines should be sited somewhere, but who refrain from the idea of having them sited in their town or neighbourhood.

9   For more details about the choice experiment referred to here, see Ek (2006), and for an introduction to choice experiments as an environmental valuation method, see Hanley et al (2001). A similar analysis of the environmental impacts of wind power in Spain can be found in Alvarez-Farizo and Hanley (2002).

10 Yet another permit, from the government, is required if the project includes three or more wind turbines with a total capacity of 10MW or more.

11 According to a European perspective, it is worth noting that nuclear opposition has often been especially effective in countries where 'a substantial devolution of authority does exist (e.g. Scandinavia and Germany), or in [countries] where the central control over a heterogeneous nation has lapsed (e.g. Italy and Spain)' (Lucas, 1981, p181). Conversely, in countries where the decision-making has been left to a small and powerful group and the opposition has little access to the political and legal systems (e.g. Belgium and France), it has been more difficult to hamper nuclear power projects effectively (see also Söderholm, 1998).

12 'Cirkulære om planlægning for og landzonetilladelse til opstilling af vindmøller (vindmøllecirkulæret)', CIR no 100, 10 June 1999 (LBK no 763, 11 September 2002), Directive on Planning and Land Zone Permission for the Establishment of Wind Turbines.

13 The Swedish government has, though, initiated a Wind Power Commission, which will consider such changes.

REFERENCES

Alvarez-Farizo, B. and Hanley, N. (2002) 'Using conjoint analysis to quantify public preferences over the environmental impacts of wind farms: An example from Spain', *Energy Policy*, vol 30, pp107–116

Åstrand, K. and Neij, L. (2006) 'An assessment of governmental wind power programmes in Sweden using a systems approach', *Energy Policy*, vol 34, pp277–296

Bärring, M., Nyström, O., Nilsson, P-A., Olsson, F., Egard, M., and Jonsson, P. (2003) *El från nya anläggningar – 2003 [Electric Power from New Plants – 2003]*, Report 03:14, Elforsk, Stockholm

Basse, E. M. (2001) 'Almindelige emner' and 'Forvaltningens efterfølgende prøvelses- og ændringsmuligheder', both in Basse, E. M. (ed) *Miljøretten, Part I*, Jurist- og økonomforbundets Forlag, København

Bengtsson, M. and Corvellec, H. (2005) *Etablering av vindkraft i Sverige, En kartläggning av miljötillstånds-processer 1994–2004*, Working Paper 2005:1, Centre for Public Sector Research (CEFOS), Gothenburg University, Sweden

Bergek, A. (2002) *Shaping and Exploiting Technological Opportunities: The Case of Renewable Energy Technology in Sweden*, PhD thesis, Department of Industrial Dynamics, Chalmers University of Technology, Gothenburg, Sweden

Bird, L., Bolinger, M., Gagliano, T., Wiser, R., Brown, M. and Parsons, B. (2005) 'Policies and market factors driving wind power development in the United States', *Energy Policy*, vol 33, pp1397–1407

Buen, J. (2003) 'Wind power developments in Denmark and Norway: A comparison', Paper presented at the workshop Strategies for Sustainable Energy Technology, Trondheim, 20–21 November

Cerveny, M. and Resch, G. (1998) *Feed-in Tariffs and Regulations Concerning Renewable Energy Electricity Generation in European Countries*, Energieverwertungsagentur (EVA), Vienna, Austria

Claeson Colpier, U. and Cornland, D. (2002) 'The economics of the combined cycle gas turbine – An experience curve analysis', *Energy Policy*, vol 30, pp309–316

Danish Wind Industry Association (1993) *Holdningsundersogelse*, Ringkjobing, Denmark

Ek, K. (2002) *Valuing the Environmental Impacts of Wind Power: A Choice Experiment Approach*, Licentiate thesis 2002:40, Division of Economics, Luleå University of Technology, Luleå, Sweden

Ek, K. (2005) 'Public and private attitudes towards "green" electricity: The case of Swedish wind power', *Energy Policy*, vol 33, pp1677–1689

Ek, K. (2006) 'Quantifying the preferences over the environmental impacts of renewable energy: The case of Swedish wind power,' in Pearce D. W. (ed) *Valuing the Environment in Developed Countries: Case Studies*, Edward Elgar Ltd, Cheltenham, UK

Ek, K. and Söderholm, P. (2005) 'Technology diffusion and innovation in the European wind power sector: An econometric analysis,' in Ek, K. (ed) *The Economics of Renewable Energy Support*, PhD thesis, Economics Unit, Luleå University of Technology, Sweden

Fisher, C. and Newell, R. (2004) *Environmental and Technology Policies for Climate Change and Renewable Energy*, Discussion Paper 04-05, Resources for the Future, Washington, DC

García-Cebrián, L. I. (2002) 'The influence of subsidies on the production process: The case of wind energy in Spain,' *The Electricity Journal*, vol 15, pp79–86

Hammarlund, K. (1997) *Attityder till vindkraft*, Occasional Paper 1997:2, Department of Human and Economic Geography, Göteborg University, Göteborg, Sweden

Hanley, N., Mourato, S. and Wright, R. E. (2001) 'Choice modelling approaches: A superior alternative for environmental evaluation?', *Journal of Economic Surveys*, vol 15, pp453–557

Hassett, K. A. and Metcalf, G. E. (1995) 'Energy tax credits and residential conservation investment: Evidence from panel data', *Journal of Public Economics*, vol 57, pp201–217

IEA (International Energy Agency) (no date) *Electricity Information*, database updated annually, OECD, Paris

Jaffe, A. B. and Stavins, R. N. (1995) 'Dynamic incentives of environmental regulations: The effects of alternative policy instruments on technology diffusion', *Journal of Environmental Economics and Management*, vol 29, pp43–63

Jaffe, A. B., Newell, R. G. and Stavins, R. N. (2000) *Technological Change and the Environment*, Discussion Paper 00-47, Resources for the Future, Washington, DC

Jaffe, A. B., Newell, R. G. and Stavins, R. N. (2005) 'A tale of two market failures: Technology and environmental policy', *Ecological Economics*, vol 54, pp164–174

Khan, J. (2003) 'Wind power planning in three Swedish municipalities', *Journal of Environmental Planning and Management*, vol 46, pp563–581

Koomey, J. G., Sanstad, A. H. and Shown, L. J. (1996) 'Energy-efficient lightning: Market data, market imperfections, and policy success', *Contemporary Economic Policy*, vol 14, pp98–111

Krohn, S. and Damborg, S. (1999) 'On public attitudes towards wind power', *Renewable Energy*, vol 9, pp945–960

Lucas, N. (1981) 'The influence of existing institutions on the European transition from oil', in Goodman, G. T., Kristoferson, L. A. and Hollander, J. M. (eds) *The European Transition from Oil: Societal Impacts and Constraints on Energy Policy*, Academic Press, London

McVeigh, J., Burtraw, D., Darmstadter, J. and Palmer, K. (2000) 'Winner, loser or innocent victim? Has renewable energy performed as expected?', *Solar Energy*, vol 68, pp237–255

Menanteau, P., Finon, D. and Lamy, M-L. (2003) 'Prices versus quantities: Choosing policies for promoting the development of renewable energy', *Energy Policy*, vol 31, pp799–812

Meyer, N. I. (2003) 'European schemes for promoting renewables in liberalised markets', *Energy Policy*, vol 31, pp665–676

Midttun, A. and Koefoed, A. L. (2003) 'Greening of electricity in Europe: Challenges and developments', *Energy Policy* vol 31, pp677–687

Pasqualetti, M. J. (2004) 'Wind power: Obstacles and opportunities', *Environment*, vol 46, pp22–38

Pedersen, E. and Persson Wayne, K. (2002) *Störningar från vindkraft: Undersökning bland människor boende i närheten av vindkraftverk*, Department of Environmental Medicine, Gothenburg University, Sweden

Pettersson, M. (2006) *Legal Preconditions for Wind Power Implementation in Sweden and Denmark*, Licentiate thesis 2006:12, Luleå University of Technology, Sweden

Reiche, D. and Bechberger, M. (2004) 'Policy differences in the promotion of renewable energies in the EU member states', *Energy Policy*, vol 32, pp843–849

Roe, B., Teisl, M. F., Levy, A. and Russell, M. (2001) 'US consumers' willingness to pay for green electricity', *Energy Policy*, vol 29, pp917–925

Rogner, H.-H. and Langlois, L. (2000) 'The economic future of nuclear power in competitive markets', in *Conference Proceedings of the 23rd Annual IAEE Conference, Energy Markets and the New Millennium*, 7–10 June, Sydney, Australia

Söderholm, P. (1998) 'Fuel choice in West European power generation since the 1960s', *OPEC Review*, vol XXII, pp201–232

Söderholm, P., Ek, K. and Pettersson, M. (2007) 'Wind power development in Sweden: Global policies and local obstacles', *Renewable and Sustainable Energy Reviews*, vol 11, pp365–400

Söderholm, P. and Klaassen, G. (forthcoming) 'Wind power in Europe: A simultaneous innovation–diffusion model', forthcoming in *Environmental and Resource Economics*

SOU 1998:152 (1998) *Vindkraften – en ren energikälla tar plats – Lägesrapport december 1998 från Vindkraftsutredningen*, Fritzes, Stockholm

SOU 1999:75 (1999) *Rätt plats för vindkraften*, Fritzes, Stockholm

Swedish Energy Agency (1998) *Vindkraft i harmoni*, Report ET 19:1998, Swedish Energy Agency, Eskilstuna

Swedish Energy Agency (2003) *Vindkraft: Fördelning av nationellt planeringsmål och kriterier för områden av riksintresse*, Report ER 16:2003, Swedish Energy Agency, Eskilstuna

Swedish Energy Agency (2004) *Konsumenten och den förnybara elen*, Report ER 12:2004, Swedish Energy Agency, Eskilstuna, Sweden

Swedish Government (2002) *Regeringens Proposition 2001/02:143 om samverkan för en trygg, effektiv och miljövänlig energiförsörjning*, Ministry of Environment and Infrastructure, Stockholm

Swedish Government (2003) *Regeringens Proposition 2002/03:40 om elcertifikat för att främja förnybara energikällor*, Stockholm

Swedish Government (2006) *Förnybar el med gröna certifikat*, Lagrådsremiss, Stockholm

Tegner Anker, H. (2001) 'Arealanvendelse, natur- og kulturebeskyttelse' and 'Planlovgivning', both in Basse, E. M. (ed) *Miljøretten, Part I*, Jurist- og økonomforbundets Forlag, København

Toke, D. (2002) 'Wind power in UK and Denmark: Can rational choice help explain different outcomes', *Environmental Politics*, vol 11, pp83–100

Wolsink, M (1996) 'Dutch wind power policy: Stagnating implementation of renewables', *Energy Policy*, vol 24, pp1079–1088

Wolsink, M. (2000) 'Wind power and the NIMBY myth: Institutional capacity and the limited significance of public support', *Renewable Energy*, vol 21, pp49–64

Zarnikau, J. (2003) 'Consumer demand for "green" power and energy efficiency', *Energy Policy*, vol 31, pp1661–1672

Chapter 8

Sharing Burdens in the European Union for the Protection of the Global Climate: A Swedish Perspective

Lena Gipperth

SHARING BURDENS

It is clear that there are considerable differences among countries as to their historic responsibility for greenhouse gas (GHG) emissions, their commitment to reducing greenhouse gases under the Kyoto Protocol (properly termed the United Nations Framework Convention on Climate Change, 1992) and their difficulties in meeting these commitments. Countries will also differ regarding the means chosen to achieve the emission targets of the Kyoto Protocol. One example is the difference in importance that states ascribe to the use of implementing mechanisms outlined in the protocol. Likewise, the legal responsibility for GHG emissions varies between different businesses and individuals, and not always in a clearly recognizable relation to actual emissions or their impacts.

This chapter focuses on how burdens to reduce GHG emissions are shared between the European Union (EU) member states and different sectors within the EU and how this policy affects the national policy in member states, particularly Sweden. The EU is responsible for about 15 per cent of the world's GHG emissions, while comprising only 5 per cent of its population (European Commission, 2002). This means that individuals in the EU are responsible for GHG emissions three times higher than the average individual on a global scale. The Kyoto Protocol commits the EU to achieving an 8 per cent reduction in GHG emissions by 2008 to 2012 compared to the 1990 level.

After a general discussion of burden differentiation, this chapter describes how the EU – in order to meet its commitment to the Kyoto Protocol – has jointly differentiated the responsibility and distributed its common reduction commitment among the member states. It outlines the impact of this distribution on member states – in particular, Sweden's ability to have its own national climate policy. After focusing on how the burdens are shared among member states, the chapter then discusses how the EU – through sector regulations and specific climate legislation – also indirectly regulates how member states are to distribute burdens for GHG emission reductions among domestic national sources. Finally, conclusions are drawn about future burden differentiation within the EU and Sweden, respectively.

Burden differentiation: A definition

In latter years, burden-sharing or burden differentiation has become a topic for scientific research and the concept has been analysed from different perspectives. One general definition describes it as 'the way in which a group of countries benefiting from a collective good agrees to share the costs of providing the collective good' (Ringius, 1999). Even if burden-sharing may be described in such general terms, the concept is closely linked to climate issues. The United Nations Climate Convention recognizes climate change as a 'common concern of humankind', but also assigns different degrees of responsibility to developed and developing countries for managing this global threat. Article 4 of the convention lists the commitments of parties according to 'their common but differentiated responsibilities and their specific national and regional development'.

The Kyoto Protocol is more explicit on this differentiated national responsibility for reducing emissions. Essentially, only developed countries have clear obligations for reducing GHG emissions; but even their responsibilities are differentiated. Several equity principles are reflected in the protocol; but the emerging differentiation scheme was not founded on a specific method. It was, rather, based on negotiations influenced by the various interests and national circumstances of the parties. The protocol gives states an opportunity to further differentiate their responsibility by taking joint actions. Article 4.1 declares that:

> *Any Parties included in Annex I that have reached an agreement to fulfil their commitments under Article 3 jointly shall be deemed to have met those commitments provided that their total combined aggregate anthropogenic carbon dioxide equivalent emissions of the greenhouse gases listed in Annex A do not exceed their assigned amounts calculated pursuant to their quantified emission limitation and reduction commitments inscribed in Annex B and in accordance with the provisions of Article 3. The respective emission level allocated to each of the Parties to the agreement shall be set out in that agreement.*

SHARING BURDENS BETWEEN EU MEMBER STATES

The EU Burden-Sharing Agreement

When ratifying the Climate Convention, the EU had already declared its intention to comply jointly. When signing the Kyoto Protocol, the EU repeated its intention to jointly fulfil the quantified emission reduction commitments under Article 4 of the protocol. The member states made their so-called called Burden-Sharing Agreement in June 1998, which distributes the total common EU commitment to achieving an 8 per cent reduction in GHG emissions by 2008 to 2012 compared to the 1990 level through the pattern of differentiated responsibility for the individual member states presented in Table 8.1.

The difference between member states' commitments under the Burden-Sharing Agreement is striking. Germany and Denmark will need to reduce their emissions by 21 per cent at the same time as countries such as Greece, Portugal and Spain may increase their emissions by 15 to 27 per cent. The agreement was implemented through a decision by the European Council in April 2002 in connection with the decision to ratify the Kyoto Protocol (Decision No 2002/358/EC). The protocol was simultaneously ratified by the EU and its member states on 31 May 2002. It should be noted that the ten new EU member states (not represented in Table 8.1) are assigned individual targets of –8 per cent (except Hungary and Poland, whose target is –6 per cent). These new member states do not participate in the internal burden-sharing for the period of 2008 to 2012. Along with the other member states (the EU 15), they do, however, have to implement EU climate policies and measures, including the EU emissions trading scheme.

Table 8.1 *Burden-sharing decided by the European Council*

Member states	Targets for 2008 to 2012 under the Kyoto Protocol and the EU Burden-Sharing Agreement (change in percentage from the 1990 level)
Austria	−13
Belgium	−7.5
Denmark	−21
Finland	0
France	0
Germany	−21
Greece	+25
Ireland	+13
Italy	−6.5
Luxembourg	−28
The Netherlands	−6
Portugal	+27
Spain	+15
Sweden	+4
UK	−12.5

Burden-sharing criteria

Neither the 1998 agreement on burden-sharing nor the 2002 implementing decision stated any explicit criteria for burden distribution. Since general equity and environmental principles may have been possible to apply, it is interesting to take a closer look at the method used for this differentiation. The general statement in the Climate Convention was, of course, relevant. It says that 'differences in Parties' starting points and approaches, economic structures and resource bases, the need to maintain strong and sustainable economic growth, available technologies and other individual circumstances, as well as the need for equitable and appropriate contributions by each of these Parties' are relevant factors to be considered when adopting national policies and taking measures to limit GHG emissions (Article 4(2)(a)). However, in a distribution situation such as the one confronting the EU, the question was not only which criteria should be considered as relevant, but also how and to what extent.

With no clear rules or criteria for distributing burdens of emission limitation, the floor was left open for negotiations. It is obvious that the 1998 agreement is a result of bargaining among the member states. During these negotiations, several factors were acknowledged as relevant. Several criteria for allocating the burdens were also discussed, based on, for example, population, gross domestic product (GDP), sectors and equal abatement costs. The distribution scheme finally agreed upon by the member states was presented by the Dutch presidency. This so-called triptych model was built on a sectoral approach to calculate reasonable emissions allowances (Phylipsen et al, 1998; Ringius, 1999).

In the triptych model, the proposed national targets are based on the sum of allowances for three categories of emissions: the power-producing sector, the export-oriented energy-intensive industry and the light domestic sectors. Different criteria are used to calculate 'reasonable' emission allowances for each of these sectors. The agreement differentiated member states' contributions according to national differences in these sectors. It furthermore acknowledged national circumstances, such as expectations for economic growth, the energy mix (e.g. amount of coal fires and potential for renewable energy), the public acceptance of nuclear energy, and industrial structure such as the state's share of heavy industry. States with an energy production based mainly on fossil fuels (such as Germany) were given a more intensive burden, whereas states with a need for further economic development (such as Greece, Ireland, Portugal and Spain) or states already having made improvements in energy efficiency (such as Sweden) were allocated less onerous burdens.

The triptych model turned out to be a useful tool in the negotiations over burden distribution among the member states. While the model was questionable in some respects, it turned out to be politically acceptable. Although there are several examples of distribution models coming out of negotiations, the EU Burden-Sharing Agreement is still a unique example of a successful negotiation on how to distribute the burdens of abating pollution.

Failure by the EU and member states to meet commitments

How, then, are the EU member states meeting their targets? An inventory made by the European Environmental Agency in 2004 showed that most member states (EU 15) are not on track with the targets of the Kyoto Protocol (EEA, 2004). With continuing trends, they are likely not to meet their commitments under the 2002 burden-sharing decision (EEA, 2005). Only four countries are ahead of their Kyoto target path (Germany, France, Sweden and the UK). Six countries are assumed to meet their targets provided that they make and take additional domestic policies and measures, whereas the other five countries are projected not to reach their targets. The latest inventory shows that total EU GHG emissions in 2003 stood at only 1.7 per cent below their 1990 level.

Given existing policies and measures, projections for the EU show total GHG emissions decreasing by 1.6 per cent between 1990 and 2010 – that is, 6.4 per cent short of the joint target of an 8 per cent reduction. Should additional planned policies and measures be implemented, this would result in a 6.8 per cent reduction. If, furthermore, the so-called Kyoto mechanisms are used, the EU target could be achieved. This would, however, rely heavily on over-compliance by several member states, something that cannot be taken for granted.

Legal aspects of member states' non-compliance

Since the EU is one of the most supportive parties behind the Kyoto Protocol, a heavy responsibility lies on the union to fulfil its commitments. The legal design of the burden-sharing decision and the legal consequences of a failure for the EU to achieve the targets and the burden distribution are thus of crucial interest.

Generally, international law would indicate that if the EU fails to achieve the 8 per cent joint reduction according to the protocol, each member state is individually responsible to meet its level of emission as assigned by the burden-sharing decision. However, Article 4(6) in the Kyoto Protocol assigns responsibility to parties only in the event of failure to achieve the 'total combined level of emissions reductions'. This means that as long as the EU, as a whole, complies with the common 8 per cent emissions reduction, both the union and its member states are seen as fulfilling all the obligations in the protocol. According to the Marrakech Agreement of 2001, however, each member state needs to individually demonstrate its reductions according to the burden-sharing decision after the commitment period. Failure by one or a few member states to comply with this decision must thus be compensated for before the expiration of the commitment period.

What happens, then, if the EU fails to meet its Kyoto target? From an internal EU perspective, it can, first of all, be noted that all EU member states are bound by council decisions. They are generally committed to taking all appropriate measures to fulfil obligations arising from the treaty establishing the European Community (EC). As shown in Table 8.1, the burden-sharing decision commits some member states to limit their emissions, while allowing others to increase their emissions, all within the common –8 per cent reduction target. A member state not meeting its reduction commitment as agreed by the council will thus be in breach of the EC Treaty and may be brought before the European Court of Justice. However, the internal EU system for handling breaches does not repair the breach of international law occurring if the common 8 per cent reduction is not achieved. It is therefore important not only to analyse the sanctions in case of a breach, but also the incentives for member states to comply or even do more than is required by the agreement.

Some member states have proposed that countries with actual or projected over-compliance – such as Sweden, the UK and Germany – should compensate for countries failing to meet reduction targets. The Swedish government argues that such solidarity would make the Swedish policy pointless and advocates the possibility of banking its reduction (Swedish Government Bill 2001/02:55, pp28, 36). However, the principle of solidarity (Article 10 of the EC Treaty) obliges member states to take all appropriate measures to ensure fulfilment of the obligations arising from the treaty. It may thus be argued that member states with excess GHG reductions are obligated to share these with member states that fail to meet their target. At the same time, however, there is no clear legal obligation for over-complying member states to share their excess achievement with states not meeting their reduction targets. Furthermore, there is currently no procedure for how member states' 'excess' GHG reductions of emissions will be handled by the EU in the case of member states failing to meet their targets. The commission has no legal instrument to force a member state to make such compensation, and furthermore lacks legal instruments to act against member states before the period for compliance has expired, even if there are clear predictions that targets will not be met.

A situation where EU compliance with the Kyoto Protocol depends upon some countries' substantial achievements is thus open to political negotiations. In such bargaining, one may question the strength of the Swedish arguments for banking if the alternative is that the EU cannot meet its joint commitments under the Kyoto Protocol. How, then, should over-compliance by some member states be compensated for? A direct transaction between over-complying and underachieving states can be performed by trading units of a state's assigned amount of GHG emissions, known as AAUs (1 assigned amount unit = 1 tonne of CO_2 equivalent emissions). Compensation for such transfer might also be a subject during negotiations of the post-2012 burden-sharing scheme. This issue may seem of merely academic relevance since the Swedish contribution to total EU GHG emissions is very small (1.7 per cent in the year 2000). However, the legal situation is similar for larger member states, such as the UK and Germany, which together contribute 40 per cent (16 and 24 per cent, respectively) to the total EU emissions (EEA, 2005).

BURDEN-SHARING WITHIN MEMBER STATES

EU Climate Policy

Under the EU burden-sharing scheme, member states are not free to choose the means to accomplish their individual undertakings. The union, to some degree, governs how member states should nationally differentiate burdens between their domestic sectors and installations. The EU frameworks for national solutions and policies will thus evidently have a strong impact on the scope of local-level involvement in climate policy. This section therefore provides an overview of the general framework surrounding EU climate policy and the regulations proposed or enacted to combat GHG emissions, looking particularly at the use of flexible mechanisms and how this legislation may affect individual member states' climate policies.

Under the EC Treaty, member states must take all appropriate measures to ensure fulfilment of the obligations arising from the treaties or resulting from action taken by the institutions of the European Community, such as legislation (Article 10, EC Treaty). Depending upon the main purpose of an EC law – for example, environmental protection or promotion of the internal market – member states have different degrees of freedom to maintain or introduce more stringent legislation than intended by a directive (Article 176, EC Treaty).

As a general rule, the EU Commission is to be directed by principle of subsidiarity (Article 5, EC Treaty), which dictates that the European Community will take action only if, and in so far as, the objectives of the proposed action cannot be sufficiently achieved by the member states and the proposed action can be better achieved by the European Community. Although new community legislation can always be questioned on grounds of this principle, member states are also required (by Article 10 of the EC Treaty) to be loyal to the EC in order to fulfil obligations arising out of action taken by the institutions of the community, such as the commitment in the Kyoto Protocol. Despite some freedom of action, member states are thus bound by EC legislation when formulating their national climate policy and selecting instruments to reduce their GHG emissions. In effect, EU legislation related to common commitments under Kyoto results in decreasing space for national policy initiatives.

More specifically, EU climate policy is based on the Sixth Environment Action Programme *Environment 2010: Our Future, Our Choice* (COM (2001) 31 final), which states the common objectives, targets and policy approach for climate change. To further develop EU policies and measures to reduce GHG emissions in a cost-effective manner, in 2000 the European Commission launched a European Climate Change Programme (ECCP) using a multi-stakeholder process (COM (2000) 88 final). The ECCP programme aims at providing the basis for the European Commission's development of legislation and other instruments. In the first phase of EECP (2000 to 2001), the focus was on investigating potential initiatives for reducing GHG emissions in the energy, transport and industry sectors. A European Commission report (ECCP,

2001) identified 42 possible measures equivalent to doubling the emission reductions required from the EU in the first commitment period of the Kyoto Protocol compared to 1990 levels. The first task for the second phase of the ECCP 1 was to implement these identified measures. Besides the EU framework directive for emissions trading (Directive 2003/87/EC), the European Commission proposed directives on the promotion of bio-fuels and combined heat and power (CHP), a voluntary commitment by car makers to reduce carbon dioxide (CO_2) emissions, as well as a communication regarding vehicle taxation.

The EU and Kyoto's flexible mechanisms

The EU's work to abate climate change is currently focused on the so-called flexible mechanisms envisaged in the Kyoto Protocol – that is the Joint Implementation (JI) and Clean Development Mechanism (CDM) measures. Agriculture is of central concern, as evidenced by investigations of the mitigation potential of the improved use and management of agricultural soil, and the potential for sequestration through afforestation and reforestation. The focus is also on the promotion of renewable energy sources in heating applications by analysing the potential for increased uptake and the ways in which both existing directives, such as the Directive on the Energy Performance of Buildings or the proposed Combined Heat and Power Directive, and new measures can contribute to promoting such applications. In October 2005, ECCP 2 was launched to provide a common policy framework for EU climate change policy beyond 2012. Areas of specific concern in ECCP 2 are transport and aviation, with an emphasis on finding new common technologies and adaptation policies.

One of the most characteristic examples of the way in which European Community legislation directly impacts upon national climate policy is the directive establishing an emissions trading scheme. Under Annex 1 of the Kyoto Protocol, parties are required to use the emissions trading mechanism as a supplement to domestic actions. The 2001 Bonn Agreement develops this further by stating that domestic action is to constitute a 'significant' element of emissions reductions. Although the EU has internationally proposed a more exact limit for the use of mechanisms, it has, so far, not set up any internal limit at the same time as several member states are dependent upon the use of flexible mechanisms, such as emissions trading, to reach their national targets. What is more, the design of the EU trading scheme directly excludes member states from using measures other than flexible mechanisms to combat emissions from certain sectors. It is also notable that Sweden is included in this EU-wide mandatory scheme despite having proclaimed that its national climate policy objective is to be reached without the use of flexible mechanisms. So, let us turn to the details of the EU trading scheme and their implications for domestic climate policies.

The EU scheme for emissions trading

In October 2003, the European Council decided on a directive establishing a scheme for GHG emissions allowance trading within the European Community (Directive 2003/87/EC). The emission trading is expected to be important in combating GHG emissions. It exemplifies how more than just environmental concerns are important when developing new EU-wide instruments since the directive is, in many ways, adjusted to avoid trade barriers and to promote the working of the common internal market.

The first preliminary phase of emissions trading covers the period of 2005 to 2007. The aim is to gain experience before the global trading scheme starts in 2008. In this first phase of emissions trading, only CO_2 and some 'core activities' are included. The scheme thus covers approximately 45 per cent of estimated EU CO_2 emissions in 2010 – equivalent to 30 per cent of projected total GHG emissions within the EU – and comprises more than 9000 installations. The categories of activities covered in the first phase include energy activities; production and processing of ferrous metals; mineral industry; and some industrial plants producing pulp, paper and board. Several member states have strongly advocated the possibility of opting certain sectors out to be able to exclude installations. The UK and Denmark already have national trading schemes and want them excluded from the common EU scheme. On the other hand, some countries have pleaded for opting sectors in. Sweden argued that this should be possible for the transport sector.

The single market for carbon dioxide includes all member states. Most of the ten new member states experienced decreasing national GHG emissions after 1990, thus enjoying the possibility of becoming net sellers under the EU trading scheme. There was, accordingly, a risk that reduction credits achieved by the stagnant economy in these countries would be sold at the EU emissions market, thus keeping the price of allowances low. So far, this risk seems to have been overestimated, and the price of allowances has generally advanced.

Allocation of trading allowances among and within the member states

The allocation of allowances to participating companies is an indirect way of distributing burdens and is therefore an issue intensively discussed within the EU. The European Commission argued for having one common allocation approach in the interest of protecting the harmonization of the internal market. Auctioning allowances in one member state while allocating them freely in another may lead to distorted competition between companies in the EU. To counter this potentially detrimental variation in national approaches, the European Council decided that allowances will be free of charge for the first trading period. The experiences of this free-of-charge approach will then be reviewed before deciding on the method of allocation for 2008 to 2012.

The trading directive requires member states to devise a national allocation plan (NAP), presenting the total number of allowances to be created for the period and the distribution of these allowances to individual plants. For the first trading period, the plan was to be sent to the European Commission before March 2004 (Annex III of the directive). In the *EU Emissions Trading Scheme: How To Develop a National Allocation Plan*, the European Commission provided guidelines on methods for drawing up such plans (European Commission, 2003). The quantity of allowances to be created is to be decided by each member state on the basis of a number of common criteria and principles set out in the directive and commented upon in the guidelines. To minimize the risk of distorting competition due to member states' different principles for initial allocation, the commission can reject an NAP that does not comply with the criteria of the directive. The European Commission is also charged with evaluating the risk that member states may allocate allowances in ways that constitute state support incompatible with (and possibly even in breach of) Article 87–89 of the EC Treaty.

Admittedly, different member states have interpreted the principles differently, not least depending upon the character of their national climate policy. Activities intended to be included in the trading scheme argue for generous allocation, while competitors not included and environmental organizations want a more restricted allocation. However, the NAPs have indirectly been an instrument for distributing burdens among activities inside and outside of the trading scheme.

Following critique from British industry over comparably insufficient allocation of allowances, the UK proposed an amendment of their NAP (which was already presented to the European Commission) that would allow for increasing the total submitted quantity of allowances in the NAP. The commission rejected the proposed amendments in April 2005, arguing that the initial UK allocation was in compliance with the criteria (Commission Decision C(2005)1081 final of 12/IV/2005). The UK then launched proceedings against the European Commission and in November 2005 the Court of First Instance (Case T-178/05) annulled the Commission's Decision. In February 2006, the Commission again decided not to consider the proposed amendments to the provisional NAP (Commission Decision C(2006)426 final of 22/II/2006), whereupon the British government decided not to pursue further court actions in order to establish certainty for the installations covered by the trading scheme. It is notable that the Commission, in its decision on the first set of NAPs for the 2008–2012 trading period, reduced the amount of allowances proposed in the NAPs by almost 7 per cent (European Commission, 2006). This indicates that the Commission now assesses the fulfilment of the criteria set up in the Trading Directive in a more rigid way.

A final cap on total allowances in each member state and totally within the EU is thus not placed until the European Commission accepts the NAPs. The Swedish cap is 22.4 million tonnes of CO_2 emissions annually for existing activities. For new installations, there is an additional allowance equivalent to 2.19 million tonnes of CO_2 for the whole period of 2005 to 2007 (Swedish Ministry of Industry, Employment and Communications, 2004). It should be noted that the 1990 baseline amount of CO_2 emission equivalents from the activities covered by the EU trading allowance scheme has been estimated at 21.1 million tonnes (SEPA and SNEA, 2004). This, of course, means that if Sweden is to reach the national climate objective of a 4 per cent reduction by 2012, this would require a much lower cap in the 2008 to 2012 phase of trading, and/or that the burden be distributed to other sectors such as transport and heating (Engelbrektson, 2002). Again, we see that the member states' room for national climate policy initiatives is, to a large extent, governed by EU policy.

The NAP is also a way of distributing burdens within the group of activities that are included in the trading scheme. Based on the principles of the trading directive, the Swedish NAP presents different methods for distributing allowances, on the one hand, between existing and new activities and, on the other, between different sectors. The Swedish NAP takes account of the degree to which emissions are linked to the use of raw materials or certain fuels. Other relevant factors include historic emissions and competition from non-European activities. As a result, the Swedish NAP distinguishes between, for example, incineration plants within the energy production sector and the iron and steel industry. While the former only receives trading allowances equivalent to 80 per cent of their existing emissions, the latter will obtain 100 per cent. The reasons are that the iron and steel industry is highly exposed to competition and has low or no potential to decrease carbon dioxide emissions, while the energy production sector enjoys the option of forwarding increased clean-up costs onto its customers.

The sectoral approach used suggests how the burdens between member states are to be distributed, and the methods used in the NAPs exemplify how rather abstract principles can be developed into more precise practical methods. Admittedly, they are exposed to criticism with respect to the resulting balance among potentially conflicting values and factors. However, in a transparent process, such approaches might be a way of further developing more general principles of fairness and equity.

EU-wide trading versus command and control

When designing a new EU-wide emission trading scheme, an essential question concerns how to link it to existing community legislation. This is, in particular, the case for directives with a more traditional command-and-control approach, such as the Directive on Integrated Pollution Prevention and Control (IPPC) (Directive 96/61/EC). This IPPC Directive introduced an integrated permit scheme for large industrial point sources, covering all pollutants, including CO_2. Member states are required to ensure that installations have a permit and apply the best available techniques to prevent different kinds of pollution. However, the trading scheme has been decided to override the IPPC permit system; installations covered by the trading scheme must not include limits on direct GHG emissions except insofar as these may have significant local effects. When allocating allowances, member states are also required to ensure that the total collective emissions from all the installations participating in the trading scheme do not exceed the level of protection that would have been required if the IPPC Directive had been applicable for these emissions.

The introduction of a trading scheme where emissions of GHG are treated separately poses challenges to member states such as Sweden and Germany with a national tradition of integrated control of environmentally hazardous activities. The trading scheme confronts the specific interest for environmental examination and control of installations with the more holistic view of using the economic potential of an activity in the environmentally best way. At issue here is whether it is really appropriate to detach one type of emission from already well-functioning national permit and control systems that are also well suited to controlling all types of emissions from large installations.

The introduction of a new instrument to combat GHG emissions from sources already controlled by a permit system could also be seen as posing – indirectly – a question on how to distribute burdens. Since the trading scheme is intended to be supplementary to the already existing permit system, the burden to decrease emissions is not to be less in relation to effect – that is, the amount of reduced emissions. What makes a difference, though, is that the trading scheme aims at lowering the costs of reducing emissions, thereby leaving economic scope for further environmental protection within the more traditional system of command and control.

Sources such as transportation have shown increasing GHG emissions and are not under the same command-and-control system as stationary sources. They are, furthermore, not yet included in the trading scheme and will be more difficult to include for both economic and political reasons (Engelbrektson, 2003). The result of introducing the emissions trading scheme for only certain relatively easily controlled activities can be seen as putting a heavier burden on such activities compared to activities hitherto excluded from the trading scheme. On the other hand, the burdens placed on activities included in the trading scheme provide the actors with a foresight of at least three years (under the NAP). The burden of other sectors currently excluded from the trading scheme do not profit from this type of foresight and thus do not enjoy the same guarantees against the risk of unforeseeable extra burdens in the not-too-distant future.

The EU and project-based mechanisms

In November 2002, a working group of the ECCP presented a report on how to link Joint Implementation and Clean Development Mechanism measures to the emissions trading system, and on how to promote the private sector to start JI/CDM projects. The report saw project-related mechanisms as a complement to emissions trading, with the potential to cover the whole European Economic Area. The European Council decided in October 2004 on a directive linking the JI and CDM project-based mechanisms to the EU emission trading scheme (Directive 2004/101/EC amending Directive 2003/87/EC). The use of project-based mechanisms aims at cost-effective measures to decrease emissions, particularly for CDM projects generating positive side effects for economic development. However, it is also arguable that by way of CDM projects, developed states get credit for relatively cheap emission reductions, thus leaving the developing countries with no other option but to use more expensive measures to reach their commitments for future decreases in GHG emissions.

Several European countries have introduced programmes to develop these mechanisms. Together with Canada and Norway and companies from some other countries, including Japan, Finland and Sweden, The Netherlands participated in the World Bank Prototype Carbon Fund (PCF). The Dutch government plans to achieve half of The Netherlands' reduction burden through projects abroad. To that effect, it has an extensive programme to help Dutch companies invest in renewable energy and energy efficiency in Central and Eastern Europe, and buy carbon credits generated from these projects (Dutch Ministry of Housing, Spatial Planning and Environment, 2002). Committed to reducing its GHGs by 21 per cent, Denmark has a programme for Joint Implementation in the Eastern European countries. Since 1993, Sweden has been involved in a series of joint projects with actors in the Baltic States and Russia to develop systems for sustainable energy supply and more efficient energy use. Conducted under the Climate Convention pilot programme for Activities Implemented Jointly (AIJ), the projects include loans, credits and competence transfer (SOU, 2002). A national commission concludes, however, that the Swedish National Energy Administration will be the main actor and that there are, so far, no incentives for private companies to get involved in JI projects (Swedish Governmental Bills 2001/2002:55 and 2001/02:143).

Many member states (EU 15) have already planned for or begun projects in the ten new EU member states. Candidate countries were required to adopt EU environmental legislation when becoming members. This generally raised their levels of environmental protection and the baseline for counting credits for joint projects, thereby reducing the additionality. The outcome in credits for the investing country may therefore not be as high as initially expected (Engelbrektson, 2002).

CONCLUSIONS

Direct and indirect burden-sharing

There is a considerable difference among states and even, sometimes, groups of states or continents in terms of historic reductions and current amounts and sources of GHG emissions. States, economic actors and activities have historically differed in their readiness to combat emissions, and their technical as well as economic options to decrease emissions have varied. States also differ with respect to their potential for reducing these emissions due to differences in industrial structures, emitting activities, etc. Furthermore, the political will to reduce emissions and to combat climate change still differs substantially among individual states.

To be able to jointly fulfil individual national commitments to the Kyoto Protocol, the EU designed a way of distributing the burdens of emissions reduction and, in so doing, of taking some account of national differences. The EU Burden-Sharing Agreement and the ensuing NAPs bear witness to this open and

direct sharing of a commonly defined burden, emanating from global negotiations and agreements. On the other hand, we have also seen examples of
indirect and somewhat less open burden-sharing. This can be found in the mix
of regulations and instruments regarding different sectors in both EU and
national policy. Other examples concern the EU decision to include certain
sectors in the trading scheme, but not others, and the degree to which project
mechanisms may be used by member states.

Developing differentiation methods

Several general principles have obviously been deemed relevant in the process
of differentiating the burdens to decrease GHG emissions among the member
states. This goes also for the NAP distribution of allowances among sectors and
activities. The EU Burden-Sharing Agreement did rely on the polluter pays
principle (PPP) as an important base for EU environmental policy, in combination with other principles of equity. Equally relevant to the responsibility for
emissions seems to have been the capacity to carry out and pay for the reduction, and the need for individuals to enjoy a decent standard of living (Ringius
et al, 2000; Torvanger and Ringius, 2000). One could argue that the PPP was,
to some degree, subsumed under values of socio-economic development and
welfare. On the other hand, one could also argue that the PPP has been developed in the climate context. By taking into account other aspects than merely
current and direct pollution, such as previous pollution and environmental
benefits (e.g. the phase-out of nuclear power), the view of the polluters' responsibility becomes broader and more complex. The need for such development is
even more urgent in the light of the impact that new members states will have
on future burden differentiation.

The differentiation between member states in the EU Burden-Sharing
Agreement, as well as the distribution of burdens in many of the NAPs, can,
however, be questioned in terms of their meeting principles of fairness and
equity. Just to mention one example: Denmark claims to count its reduction
target from a baseline that is 5 million tonnes of CO_2 higher than the actual
emissions in the baseline year of 1990, arguing that this is necessary in order to
compensate for the unusually high Danish import of electricity that year. Future
burden-sharing agreements will most certainly not be founded on the same
criteria as the first agreement. And even if the next generation of NAPs are
based on the same criteria, these criteria will most certainly be interpreted
differently.

The differentiation methods described above may be seen as an example of
how principles of equity may be developed and may inspire further development in other contexts – for example, on the international level, but also on a
national level where burden-sharing is indirect. Indeed, the methods for finding
politically accepted decisions have been further developed and are now based
on more sophisticated criteria (Jansen et al, 2001; Berk et al, 2003). This development can be expected to continue as the recognition and acceptance of the
need for further reductions increase. The use of explicit models and criteria for

equity as a base for political negotiations may thereby increase the transparency of how burdens are distributed. One should not forget, however, that models and criteria are only decision support tools, and they do not replace the need for political negotiations.

Do weak enforcement tools jeopardize the distribution of burdens?

Decisions on the distribution of burdens among states, sectors and activities depend upon the level of trust that parties hold for each other. The lack of functioning and effective enforcement tools might damage future possibilities of distributing burdens and deal blows to the level of trust. As it now stands, the EU's legal framework for burden-sharing provides member states with low incentives to go further or faster than agreed upon. The Swedish example shows that there is a clear risk involved here: over-compliance by one member state may be used to compensate non-compliance by other states in order to achieve the EU's overall global commitment. Even if member states can uphold the right to some kind of compensation, over-compliance is risky business. From the perspective of trust, the European Environment Agency's conclusion that the EU, as a whole, needs some states to over-comply to be able to compensate for other states' non-compliance is therefore alarming. There is an urgent need for clarification on how to handle this situation so that member states can trust each others' readiness and actions to achieve their targets, and not risk becoming victims if they over-comply.

As long as the path to fulfilling the burden-sharing decision is not directly based on the binding articles of the EC Treaty, the EU Commission has no legal power to enforce stronger actions and must thus resort mainly to political means to influence member states' national climate policies. At the same time, one should note that a projected trend of not meeting the target is not a breach of the EC Treaty as long as common legislation is fully implemented. The legal instruments to force member states to meet their targets can only be used after it is shown that the targets are not met, and not just on the basis of predictions.

A strong – and shared – responsibility

The need to reduce GHG emissions is a gigantic task and a common responsibility confronting governments at all levels, from the local and national all the way to the global. At a regional European level, the EU has been able to stabilize emissions. However, the recent increase of emissions in some sectors and states is alarming. On the other hand, there is also great potential to meet targets, at least in the first periods. The 2001 ECCP report found more than 40 measures and policies with a potential of reducing emissions twice as much as the current EU commitment. The report also indicates that the EU emissions trading scheme has a potential of reducing CO_2 emissions by 700 million tonnes, twice as much as required by the Kyoto Protocol.

The EU is a major party in promoting the Kyoto Protocol, and the expectations on the EU to achieve its targets are thus quite high. A failure by the union to jointly meet its commitment will have negative impacts, not only on the reliability of the EU's climate policy, but also on the development of the international climate regime. Furthermore, it might turn out to have a profound impact on the level of trust and the smoothness of cooperation among the member states. In view of the many often diverging sector interests in the EU, the union constantly finds itself in a situation of balancing the longer-term need of reducing greenhouse gases against more short-term interests pressing for economic development and specific favours. Development of the EU climate policy is also a question of how to balance the common interest versus national interests, and on what level policy initiatives are to be taken. The development of a common climate policy within the EU so far indicates decreasing authority and room for decision-making at national and local levels. At the same time, the EU is clearly dependent upon national and local actions to reduce GHG emissions.

To jointly achieve the common target set up in the Kyoto Protocol thus presupposes a strong political readiness for close collaboration between member states, and a high level of trust in each others' intentions, as well as a supportive legal framework and effective instruments. As the EU expands and obtains new members, future burden-sharing agreements are likely to become an even greater challenge. An expanded EU will not only involve more parties in the negotiations. It will also mean a larger burden to share, more factors to consider and a more complex political situation, with correspondingly complicated negotiations. One should note that the ten 'new' member states have decreased their emissions considerably more than their commitment under the Kyoto Protocol. How this should be considered in a future burden-sharing agreement is currently neither openly discussed nor possible to foresee. A further development of methods linking the choices of distributive mechanisms to general principles might be a tool to facilitate such negotiations.

REFERENCES

Berk, M. M., Gupta, J. and Jansen, J. C. (2003) 'Comprehensive approaches to differentiation of future climate commitments – some options compared', in van Ierland, E. C., Gupta, J. and Kok, M. T. J. (eds) *Issues in International Climate Policy: Theory and Policy*, Edward Elgar, Cheltenham, UK

Commission Decision C(2005) 1081 final of 12/IV/2005 Concerning the Proposed Amendment to the National Allocation Plan for the Allocation of Greenhouse Gas Emission Allowances Notified by the United Kingdom in Accordance with Directive 2003/87/EC of the European Parliament and of the Council, http://europa.eu.int/comm/environment/climat/pdf/uk_bis_final_en.pdf

Commission Decision C(2006) 426 final of 22/II/2006 Concerning the Proposed Amendment to the National Allocation Plan for the Allocation of Greenhouse Gas Emission Allowances Notified by the United Kingdom in Accordance with Directive 2003/87/EC of the European Parliament and of the Council, http://europa.eu/environment/climat/pdf/uk_final_3_en.pdf

COM (2000)88 final *Communication from the Commission to the Council and the European Parliament on EU Policies and Measures to Reduce Greenhouse Gas Emissions: Towards a European Climate Change Programme*

COM (2001) 31 final, *Environment 2010: Our Future, Our Choice*, the sixth Environment Action Programme 2001–10, European Commission, Brussels

Decision No 2002/358/EC of the Council of the European Union of 25 April 2002 Concerning the Approval, on Behalf of the European Community, of the Kyoto Protocol to the United Nations Framework Convention on Climate Change and the Joint Fulfilment of Commitments thereunder

Directive 96/61/EC Concerning Integrated Pollution Prevention and Control (the IPPC Directive)

Directive 2003/87/EC of the European Parliament and of the Council of 13 October 2003 Establishing a Scheme for Greenhouse Gas Emission Allowance Trading within the Community and Amending Council Directive 96/61/EC

Directive 2004/101/EC of the European Parliament and of the Council of 27 October 2004 amending directive 2003/87/EC establishing a scheme for greenhouse gas emission trading within the Community, in respect of the Kyoto Protocol's project mechanisms

Dutch Ministry of Housing, Spatial Planning and Environment (2002) 'The progress of The Netherlands climate change policy: An assessment at the 2002 evaluation moment', www2.minvrom.nl/Docs/internationaal/evaluation_note_climate.pdf

ECCP (2001) *European Climate Change Report*, European Climate Change Programme, June, www.europa.eu.int/comm/environment/climat/pdf/eccp_longreport_0106.pdf

EEA (European Environmental Agency) (2004) *Annual European Community Greenhouse Gas Inventory 1990–2002 and Inventory Report 2004*, Technical Report No 2/2004, http://reports.eea.europa.eu.int/technical_report_2004_2/en/Tech_2_2004_GHG_inventory_draft.pdf

EEA (2005) *Greenhouse Gas Emission Trends and Projections 2005*, EEA Report No 8/2005, Copenhagen

Engelbrektson, I. (2002) 'Joint implementation in accession countries', in Appendix 2 to *Swedish National Inquiry Report*, SOU 2002:114, *Gemensamt genomförande – avtal för bättre klimat*, Swedish Government Office, Stockholm

Engelbrektson, I. (2003) 'EG:s begränsning av handel med utsläppsrätter – transportsektorn som exempel', in Johansson, S. O. (ed) *Nya och gamla perspektiv på tranporträtten*, Swedish Maritime Law Association, Skrifter 78, Sweden

European Commission (2002) *EU Focus on Climate Change*, Report from Directorate-General for the Environment, www.europa.eu.int/comm/environment/climat/pdf/climate_focus_en.pdf

European Commission (2003) *The EU Emissions Trading Scheme: How to Develop a National Allocation Plan – The Commission Gives Guidelines on Methods for Drawing Up Allocation Plans*, www.europa.eu.int/comm/environment/climat/030401nonpaper.pdf

European Commission (2006) IP/06/1650, 29 November, available for download at http://europa.eu/rapid/pressReleasesAction.do?reference=IP/06/1650&format=HTML&aged=0&language=EN&guiLanguage=en

Jansen, J. C., Battjes, J. J., Sijm, J. P. M., Volkers, C. H., Yberma, J. R., Ormel, F., Torvanger, A., Ringius, L. and Underdal, A. (2001) *Sharing the Burden of Greenhouse Gas Mitigation*, Final Report of the Joint CICERO-ECN Project on Global Differentiation of Emission Mitigation Targets among Countries, CICERO Working Paper, Oslo, Norway

Phylipsen, G. J. M., Bode, J. W., Blok, K., Merkus, H. and Metz, B. (1998) 'A Triptych sectoral approach to burden differentiation; GHG emissions in the European bubble', *Energy Policy*, vol 26, pp917–979

Ringius, L. (1999) 'Differentiation, leaders, and fairness: Negotiating climate commitments in the European Community', *International Negotiation*, vol 4, no 2, pp133–166

Ringius, L., Torvanger, A. and Underdal, A. (2000) *Burden Differentiation of Greenhouse Gas Abatement: Fairness Principles and Proposals*, ECN Report ECN-C-00-011 and CICERO Working Paper 13 CICERO, Oslo, Norway

SEPA (Swedish Environmental Protection Agency) and SNEA (Swedish National Energy Agency) (2004) *Kontrollstation 2004: Naturvårdsverket och Energimyndighetens underlag till utvärdering av Sveriges klimatstrategi*, www.naturvardsverket.se/dokument/klimat/pdf/huvud.pdf

SOU 2002:114 (2002) *Gemensamt genomförande – avtal för bättre klimat*, Swedish Government Office, Stockholm

Swedish Governmental Bill 2001/02:55 (2001a) *Swedish Climate Strategy*, Swedish Government Office, Stockholm

Swedish Governmental Bill 2001/02:143 (2001b) *Cooperation for a Safe, Efficient and Environmentally Sound Energy Distribution*, Swedish Government Office, Stockholm

Swedish Ministry of Industry, Employment and Communications (2004) *Swedish National Allocation Plan*, 22 April 2004, www.regeringen.se/content/1/c6/01/90/18/e9286dc2.pdf, accessed 10 April 2006

Torvanger, A. and Ringius, L. (2000) *Burden Differentiation: Criteria for Evaluation and Development of Burden Sharing Rules*, CICERO Working Ppaper 2000:1, ECN-C-00-013, Cicero, Oslo, Norway

9

Governing the Climate, (B)ordering the World

Johannes Stripple

INTRODUCTION

Much thinking on the environment is based upon the idea of the boundary. For example, degradations linked to soils, forests and freshwater ecosystems are local problems because they are 'within' a certain boundary. Acidification is a case of transboundary pollution, and stratospheric ozone depletion and the climate issue are global environmental problems. Hence, the supposed location of a certain issue within a political geography designates its character.

The discipline of international relations became interested in the environment only when it was 'discovered' that the scale of environmental issues transcended national borders. Conca (2006) summarizes 20 years of international environmental politics as a paradigm driven by the idea of 'pollution beyond borders'. It is the predominant concept of politics as territorial – that is, a space with firm borders – that has enabled the paradigm. Any analysis aiming beyond that paradigm needs first to come to terms with its territorial point of departure.

Within the overall theme of this book – which is about exploring contemporary tensions in multilevel climate governance – the specific contribution of this chapter is to provide a reflection on climate change policy-making in relation to territoriality. The chapter illustrates how territoriality – the idea that political authority is spatially organized and marked by clear boundaries – continues to shape how the international community understands and manages the climate issue. The chapter outlines two cases that illustrate the salience of territoriality in defining the contours of climate governance: the inclusion of terrestrial 'sinks' in the climate negotiations, and the debate concerning upon

137

what basis emissions of carbon dioxide should be accounted (country versus per capita).

With the sink issue, the chapter first outlines the articulation of the carbon cycle in global terms, and then the rearticulation of the carbon cycle as 'sinks' on territorial ground. It sketches the main scientific and political developments that are associated with the transformation of the global carbon cycle into 'national sinks'. This transformation is not surprising as it fits into a historic lineage of Western political imagination that makes pieces of the environment coincide with the sovereign state system in the form of 'natural resources'. The sink issue also illustrates practices of environmental governance that might be called 'imperial' – that is, understood both as reterritorialized control over the South by the North and as the establishment of an epistemological empire of carbon management and control.

With the emissions allocation issue, I first note the difficulties that international relations has had in relating to ethical considerations because of the way a certain ethical standpoint is already implicated in that discipline. I then briefly reconsider the debate on how to account for emission rights, and on what basis the initial distribution ought to be allocated. In my view, it turns out that justice has, in this context, been a predominantly nationalist discourse – that is, territoriality as defining for political community seems largely to be reconfirmed.

These two cases illustrate that current practices of climate governance are not just simply unfolding through different administrative levels and crossing borders of political authority. Instead of repeating the worn and torn question of 'how to govern the climate in a bordered world', this chapter calls attention to the possibility that governing the climate in fact contributes to a bordering and, hence, ordering of the world.

SCALES, LEVELS, BORDERS AND THE ENVIRONMENT

The most important form of contemporary political authority is usually understood to be rule over space – that is, territoriality. This is visible in the influential idea of a spatial 'mismatch' between the environmental and the political. The integrated, complex and interdependent web of the world's ecosystems is seen to be confronted with the fragmented international system of separate jurisdictions. The opening phrase of the 1987 World Commission on Environment and Development stated that 'The Earth is one, but the world is not', and later on the report states that:

> From space, we see a small and fragile ball dominated not by human activity and edifice, but by a pattern of clouds, oceans, greenery and soils. Humanity's inability to fit its activities into that pattern is changing planetary systems fundamentally. (WCED, 1987, p308)

Climate change has received almost iconic status as one of the issues on the new ecologically interdependent and truly global agenda. The atmosphere is shared by everyone and the impacts of climate change can be felt everywhere regardless of their origin. To govern the climate in a bordered world is thus put forward as *the* crucial problematic and a main challenge for climate policy-making by many different literatures. It is a starting point for writings within international relations on such themes as regimes, multilateral diplomacy and common pool resources, and it is *the* predicament to be overcome by much of the writings within green political theory on democracy, the ecological state and environmental citizenship. The common denominator is that the climate issue enters a specifically organized world and produces a certain challenge to that world.

In recent years, an important trend in geography has been a general asser-tion on the social production of space (e.g. Soja, 1989; Lefebvre, 1991; Ó Tuathail, 1996). It is in this sense that we could understand territoriality as a socially produced categorization of people and things by their location in space (Paasi, 2003). In Sack's analysis, territoriality emerges as a strategy that humans employ to control people and things by controlling an area (Sack, 1986). Historically, the territorial division of the world into independent states led to the development of the modern system of states. With decolonization, the process of carving up inhabitable land-space into discrete and exclusive sover-eign areas was greatly accelerated, and by the mid 20th century all terrestrial space was claimed as territory.

This development has a tremendous influence on the discipline of interna-tional relations (IR), which has taken this conception of bounded space as its point of departure. For IR, the state has been an entity that 'possesses a govern-ment and asserts sovereignty in relation to a particular portion of the Earth's surface and a particular segment of the human population' (Bull, 1977, p8). It has also had a massive influence in how nature, 'the environment', has been managed and acted upon:

> *The modern sovereign state is a particular political construction for which the environments do not come ready made. The task of moulding environments to fit the sovereign state is that of govern-ment.* (Kuehls, 1998, p49)

Recent literature in IR and political geography has argued that our world of sovereign territoriality is not a fact of nature, a bloodless principle of interna-tional law, but a human artefact and a highly variable practice (see Biersteker and Weber, 1996; Krasner, 1999; Caporaso, 2000; Biersteker, 2002). Crucially, our territorially organized world must be regularly reproduced since 'sover-eignty is something that has to be practised through "marking" space by boundaries of various kinds – and by mapping these boundaries in an exact science' (Albert, 1999). This marking of space is visible in the two cases outlined in this chapter and illustrates that current patterns of climate gover-nance, albeit multilevel, still build upon and reproduce territoriality.

THE CARBON CYCLE AS POLITICAL SPACE[1]

Can trees save us from global warming? This question is not so naive as may seem at first sight. The fact that vegetation and soils absorb carbon has been well known for a long time; but in recent years the inclusion of this idea in the climate convention has been the subject of intense negotiations. The oceans and the terrestrial biosphere act as large 'sinks' for carbon that can be released to the atmosphere through many different land-use practices. The natural state is that carbon flows in a global carbon cycle between the atmosphere, the oceans and the terrestrial biosphere.

What complicates this ecological geography is the political logic that enters these spaces. While the ocean is an 'outside space' owned by nobody, a *mare liberum* as Hugo Grotius wrote in 1608, terrestrial land is carved up and claimed by particular states. The geopolitical claim that follows is, for example, that Swedish forests are Swedish sinks. Despite the globality of the carbon cycle, a Swedish forest that absorbs carbon should thus be counted as a national and not as a global 'sink'. The next step would be to claim that with its large areas of forest, Sweden does not make a net contribution to the concentrations of carbon in the atmosphere. Hence, Sweden does not need to restrict or reduce its consumption of fossil fuels. In this sense, trees save those who happen to be born and living on Swedish territory not from the impacts of global warming, but from having to do something to prevent it. In contrast to this, Agarwal and Narain (1991) argued in a well-known article for the need to apportion the Earth's ability to absorb carbon to each country in proportion to its share of the world's population.

The flows of carbon through the Earth's system do not easily lend themselves to the spatiality of the state system. Hence, the moulding of the carbon cycle onto territorial ground has only recently been made possible through the development of a range of techniques for controlling, modelling and measuring the biosphere. The territorialization of the carbon cycle can only be comprehensible with reference to the climate negotiations. The following sections try to point to crucial movements in the scientific and political history of the 'sink' issue in order to trace the transformation of the global carbon cycle into 'national sinks'.

The Swedish scientist Svante Arrhenius is famous for his early calculations on human-induced climate change (Arrhenius, 1896). Arrhenius understood that carbon flows throughout the Earth, and he made reference to the exchange of 'carbonic acid' between three major carbon reservoirs: the atmosphere, the terrestrial systems and the oceans (Rodhe et al, 1997). Historically, the carbon cycle has been considerably disturbed by humans through the agrarian expansion in the Northern Hemisphere, through deforestation and, today, through the massive use of fossil fuels. Arrhenius was one of the first to show that there exists a physical limit to the amount of emissions that the atmosphere can absorb without suffering serious damage and, thus, causing changes in the global climate (Rodhe et al, 1997). Today, geoscientists have constructed detailed carbon flux models that measure the amount of carbon stored and exchanged between the atmosphere, the ocean, the terrestrial reservoir and the geologic reservoir.

The climatic importance of the global carbon cycle was confirmed in the 1950s when the first computed carbon cycle models were developed. These models were primarily based on a simple three-box system containing the atmosphere, the surface layer of the ocean and the deep-sea reservoirs. Served with data on atmospheric carbon dioxide (CO_2) levels from the newly installed measurement stations on Mauna Loa and the South Pole, the carbon cycle models aimed at predicting the extent to which the increasing atmospheric CO_2 emissions from fossil fuel combustion would be buffered by ocean carbon uptake on timescales from ten to several hundred years (Bolin, 1981). During the 1970s, the terrestrial biota and soils were included in these models in order to obtain general features of the interplay between the atmosphere, oceans and land. At this time, studies of tropical forests showed that deforestation contributed to an estimated 20 to 30 per cent of average annual CO_2 emissions (Houghton et al, 1985).

Negotiating sinks

In 1991, the United Nations General Assembly urged its members to start negotiations to establish a convention regulating anthropogenic greenhouse gas emissions. Rather than founding one 'law of the atmosphere' that would embody the issue of acid rain, as well as that of the stratospheric ozone layer, the negotiations resulted in separate conventions for each issue area (Soroos, 1997). The negotiations began in February 1991, and the resulting United Nations Framework Convention on Climate Change (UNFCCC) was signed at the United Nations Conference on Environment and Development (UNCED) in Rio de Janeiro in June 1992.

The objective of the convention is the 'stabilization of greenhouse gas concentrations in the atmosphere at a level that would prevent dangerous anthropogenic interference with the climate system' (UNFCCC, 1992). The convention frames anthropogenic climate change as a global problem and a common concern for the signing parties (the states). The convention stresses the historic responsibility of industrialized countries to take the lead in climate mitigation, but also asserts that all signatory parties share the responsibility to prevent anthropogenic emissions of greenhouse gases. This global framing can be seen to partly rest on the scientific representation of the climate issue at the time of the convention.

When the UNFCCC was negotiated, the scientific interest in global flows of carbon was prevalent. However, the message that large land areas in the Northern Hemisphere sequestered atmospheric carbon and, hence, counteracted the anthropogenic greenhouse effect had gained political recognition. Leaning on the assessment of carbon cycle science in the first assessment report of the Intergovernmental Panel on Climate Change (IPCC) published in 1990, the UNFCCC acknowledges the important role that terrestrial ecosystems play in the climate system. Vegetation and soils, acting as sinks and reservoirs mainly for CO_2, largely affect the convention's objective. Consequently, the UNFCCC encourages, in Article 4.1, its signatories to make inventories of domestic sinks of greenhouse gases and to cooperate internationally to ensure that they are enhanced and conservatively managed. Notably, the uptake of carbon dioxide in the oceans is not mentioned as a 'sink' in the convention, even though the uptake of the oceans is considerably larger. The political call for inventories of national carbon uptake contributed to a substantial shift in the focus of carbon cycle science. Exit the global carbon cycle, enter the territorial sink.

Territorializing the carbon cycle

While carbon cycle science, until the signing of the Climate Convention in 1992, had been focused on carbon flows on a hemispheric or, at best, continental scale, new research was now initiated to estimate terrestrial carbon uptake within state borders. Countries with good forestry and agricultural statistics were in a strong position to accomplish this, while other countries had to develop new methods of accounting. Since carbon cycle science at this stage could not meet the political requirements for territorially based results, few parties to the convention made any efforts to present national carbon uptake in terrestrial ecosystems in the first two national communications submitted to the UNFCCC secretariat in 1994 and 1997.

Terrestrial carbon sinks entered the negotiating table in time for the Third Conference of the Parties (COP) to the Climate Convention in Kyoto in 1997. Even though the idea of sinks had been around for quite a long time:

> ... sinks only became the focus of intensive and high-level political debate in the closing stages of the protocol negotiations, as negotiators realized just how much was at stake. (Grubb et al, 1999)

The US and others argued that removal of atmospheric CO_2 by terrestrial ecosystems would not only offer the same climatic effect as traditional emission reductions, but also do so at significantly lower societal costs. By actively enhancing the natural sequestration in biomass and soils within national borders, the Annex I parties – the industrialized states that have signed the climate convention – could both counteract the build-up of atmospheric carbon and buy time for the development of low-emission technologies (Anderson et al, 2001). Towards the end of the Kyoto negotiations, the US and its allies in the so-called Umbrella Group (notably Canada, Japan, Russia and Norway) portrayed terrestrial carbon sinks as one of the keys to a cost-effective implementation of the Kyoto targets and a prerequisite for a final agreement (Grubb et al, 1999).

The inclusion of sinks in the negotiations at Kyoto was also furthered by the architecture of the Kyoto Protocol. To cut a long story short, it became evident at the time of the negotiations in Kyoto that to reach an agreement, the inclusion of various policies and measures that could increase the cost-effectiveness of the agreement was necessary (see, for example, Jacoby et al, 1998; Rowlands,

2001; Begg, 2002). The Kyoto Protocol implies that most of the developed countries should have decreased their emissions of greenhouse gases by 5.2 per cent, on average, by 2008 to 2012.

The protocol includes three political instruments or mechanisms that aim at increasing the cost-effectiveness of greenhouse gas reductions of the developed world. The first mechanism is 'emissions trading', which gives countries emitting less than has been agreed upon the opportunity to trade what remains in their 'emissions account'. The remaining two mechanisms are Joint Implementation (JI) and the Clean Development Mechanism (CDM). The JI mechanism is supposed to work as follows. If a country such as Sweden agrees to reduce carbon dioxide levels by 500,000 tonnes, it might turn out to be more cost-effective to achieve this through action in Poland. Hence, Sweden pays for investments made in Poland while setting up the decreased emissions on its own national emissions account, and thereby strengthening that account. CDM differs from JI in that CDM projects are carried out in developing countries. The Kyoto mechanisms, particularly the CDM, provide a political vehicle for the inclusion of sinks in the climate negotiations.

Article 3.3 in the Kyoto Protocol allows Annex 1 parties (i.e. developed countries) to account for removals by sinks 'resulting from direct human-induced land-use change and forestry activities, limited to afforestation, reforestation and deforestation since 1990' (United Nations, 1997). Article 3.4 also paves the way for an inclusion of sinks resulting from 'additional human-induced activities' in a later stage of the negotiations. In order to prepare for an accounting system based on 'national sinks', the Kyoto Protocol furthermore commits all Annex I parties to install a national system for measuring anthropogenic sources and sinks of greenhouse gases no later than one year prior to the start of the commitment period (i.e. in 2007). Already when the Kyoto Protocol was signed, it was clear that the sink provisions had introduced a range of technical and scientific problems. There was, for example, no agreed definition of what should be counted as a 'forest'. Furthermore, there was a lack of credible methods for accounting for terrestrial national carbon removals, and, in particular, to differentiate between the naturally ongoing carbon sequestration in biomass and soils and carbon uptake directly caused by deliberate human land-use strategies (Anderson et al, 2001).

National terrestrial sinks have been a very contentious issue since the days of Kyoto. The Conference of the Parties negotiations under the UNFCCC continued at COP 4 in Buenos Aires (1998) and at COP 5 in Bonn (1999), but collapsed at COP 6 in The Hague (2000). The breakdown of the climate talks in The Hague had many reasons (see, for example, Grubb and Yamin, 2001; Ott, 2001); but carbon sinks were clearly one of the most difficult issues to solve (Dessai, 2001; Sedjo et al, 2001). COP 6 was resumed in Bonn in 2001, and this time the negotiating parties were put under pressure to reach an agreement since the new US administration led by President George W. Bush had announced that the US would abandon the Kyoto Protocol.

The Bonn compromise was built on a system of caps and discounts that would restrict the amounts of credits gained under Articles 3.3 and 3.4 and, hence, 'manage' the large methodological uncertainty associated with the accounting of carbon accumulation in agricultural soils, grazing lands and harvested forests (Schultze et al, 2002). The Bonn agreement was consolidated at COP 7 in Marrakech later the same year. Together, these two meetings mark the end of the political phase of the sink negotiations.

Alongside the political negotiation process, there has been a tremendous development in scientific research and measurement techniques to meet the political requirements to control and manage carbon flows within state boundaries. These technologies have reinforced the representation of the 'national carbon sink' and, hence, given the territorialization of global carbon flows continued legitimacy. While large parts of the scientific community continue to investigate the global features and processes of the carbon cycle, much of contemporary carbon cycle science has conformed to the state-based definition of terrestrial sinks that prevails in the climate negotiations.

Sinks: The international and the imperial

Nature, forests and trees never just *are*. They can be raw materials, tourist destinations, wilderness, local habitats, etc. It depends upon which perspective, which source of value, has priority (Benton and Redclift, 1994; Walker, 2002b). The transformation of the global carbon cycle into territorial sinks is indicative of modern practices of governmentality in that it moulds the environment to fit into the sovereign state system (Kuehls, 1998, p49). In his recent book on transnational water governance, Ken Conca makes the general argument applicable in this case: 'The dominant understanding pervading the regime-building enterprise is that most of nature will, in fact, sit still as territory' (Conca, 2006, p49).

The discourse on carbon sinks works on the premise that biosphere space can be brought 'inside'. The space inside the state is amenable to the discourses of modernization and development. This can be seen in contrast to the outside spaces, such as ocean space, that has largely been 'constructed as a "non-territory", an untameable space that resists "filling" or "development"' (Steinberg, 2001, p34). But, of course, the forests are not empty:

> Until quite recently, vast tracts of the Amazon rain forest were considered 'empty' by the Brazilian government. Despite the fact that tens of thousands of people lived in these forests, these areas were deemed empty due to the absence of any signs of land occupation. (Kuehls, 1996)

The inclusion of sinks in the Kyoto Protocol signals that terrestrial sinks can be considered as CDM projects within the framework of the Kyoto Protocol. If Sweden or a Swedish company plants trees somewhere in the South, these trees will absorb carbon dioxide as long as they grow, and the carbon will be stored in them until they are cut down. This implies that the place where the trees are located needs to be managed, protected and supervised over a long time. How responsibility and authority over these spaces should be arranged is intensely debated in the climate negotiations (Noble and Scholes, 2001; Begg, 2002). There is also a belief that sink projects will constitute a new form of eco-imperialism (Shiva, 1994). In Dalby's words: 'One has to ask whether the South simply becomes a tree growing zone for large Northern states and corporations trying to find a way to "sink" their emissions' (Dalby, 2000).

The sink issue illustrates practices that perhaps have more affinity with the imperial than with the international. The advantage of 'retrieving the imperial' is that it is 'one of the principal routes out of the "territorial trap" contained in the idea of a sovereign state system: the notion that borders are relatively imper-meable containers of social relations' (Barkawi and Laffey, 2002). The imperial recovers the international relations of hierarchy, the peripheral and the subal-tern (Barkawi and Laffey, 2002). There are two sides of this argument, which invoke both recent debates on the 'new imperialism' in IR (e.g. Cox, 2004) and debates emanating from Hardt and Negri's (2000) 'Empire' thesis. Writings on the new imperialism are mostly about debating US power and the extent to which world order is imperial rather than hegemonic (e.g. Ikenberry, 2004), while Empire is the claim that the world is, indeed, imperial, although it is not an American empire, but nevertheless a single logic of rule.

The management of sinks through the CDM in the Kyoto Protocol opens up reterritorialized control of the South by the North. The responsibility and authority over the 'sink spaces' (e.g. plantations and management of trees) are far from clear. In this sense, areas of carbon sinks bear a familiar resemblance to other cases of control over distant subordinate places, such as 'debt-for-nature swaps' (Lafferreiere, 1994), military bases (Johnson, 2004), tourist resorts (Gossling, 2002), economic processing zones (Abbott, 1997) or by 'supporting bio-prospecting in biodiversity rich, post-colonial territories' (Eckersley, 2004). In this vein, Dalby (2004) notes that environmentalists 'have frequently promoted the establishment of protected spaces, parks and the control of populations in manners that nonetheless replicate the practices of empire'. Of course, this differs from earlier British and French territorializations in that the official politics is not imperial, but the similarities are still there.

In Fogel's (2004) analysis, the imperial argument acquires another twist. Holding advisory scientists within the IPCC and other climate-related institu-tions accountable for the idea of forests as 'empty' and available space, she cites US Department of Energy experts who have called for 'the intensive manage-ment and/or manipulation of a significant fraction of the globe's biomass' (Fogel, 2004). Fogel sees the emerging culture of carbon management as contributing to a mechanistic 'global gaze' that moves to standardize and enrol both people and the natural world into largely inaccessible global institutions. Fogel's view on the emerging sink discourse is reminiscent of the way in which Hardt and Negri (2000) have captured 'governmentalities' under the age of the fragmented, fluid and foundation-less Empire. As Dalby (2004) summarizes the condition:

> *Sovereignty is bleeding away from states in some amorphous series of rules, regulations and shared procedures that exceeds the mandates of states and sets the terms for incorporation of many*

145

institutions and peoples into an amorphous but powerful arrangement they simply term 'Empire'.

An imperial rendering of the sink issue highlights that an emergent global culture of carbon management puts the world's entire biosphere under one type of rule. While still fragmented and fluid, this 'epistemological empire' has the potential to order the world so that the Earth becomes one undivided territory. While it may be a rather gloomy and far-fetched idea, the advantage of an imperial point of departure is that it provides a distinctive way of thinking about multilevel climate governance in relation to carbon sinks. To put it bluntly, emergent structures of knowledge, power and authority in relation to the sink issue might be easier to comprehend from an imperial, rather than from an international, point of view.

TERRITORIALITY, INEQUALITY AND CO_2 EMISSIONS

The human influence on the global climate raises a range of ethical considerations. For example, how much change in climate-related parameters should be tolerated? Is the lack of scientific certainty a legitimate reason for inaction? Does cost-benefit analysis provide a fair grounding for decisions on climate policy? Do developed countries have special responsibilities to act before the poorer nations (Brown, 2001)? The climate negotiations have, in particular, brought to the fore debates on distributive justice – for example, about:

- allocating the costs of preventing climate change;
- allocating the costs of coping with the consequences of climate change;
- what a fair bargaining process would be like; and
- what a fair allocation of emissions of greenhouse gases would be (Paterson, 2001).

There are many attempts to cover these vast normative territories (Toth, 1999; Rosa and Munasinghe, 2003). Large parts of the contributions to these debates argue for the need and relevance of a specific ethical consideration or the particular merits of a certain approach (see, for example, Rowlands, 1997; Ikeme, 2003; Tonn, 2003).

The problem with most of the above-mentioned contributions is that they take for granted the ethical issues inherent in the way that international relations are organized. The fundamental question of 'how we might live' within the context of world politics (Booth et al, 2000) has already been given a crucial answer by the practice and discipline of international relations. We know this answer fairly well. It is about the unspoken necessity and naturalness of living in territorially defined political communities. However, one has to realize that international relations pose a certain challenge for reflecting on ethical considerations. The problem of inequality is already deeply inscribed in our modern accounts of the international (Walker, 2002a). As Franke (2000) states:

> *... the traditions of international thought and practices are founded upon basic Western stories regarding how humans, necessarily, come to form political society within states and, furthermore, how these territorial communities themselves must necessarily engage one another in a disordered 'social' sphere.*

Walker and Franke are highlighting the important point that international relations cannot be understood as merely an empirical arena for making claims about what is equal and unequal because the practice of international relations already implies a specific framing of these issues. Therefore, my ambition in this section is not to engage in another normative argument about justice in relation to the allocation of greenhouse gas reductions, but to explore how the idea of territoriality informs and shapes these normative debates and arguments.

Reconsidering the per capita allocation

The patterns of carbon dioxide emissions vary considerably around the world. The global mean level of carbon per person is about 1 metric tonne per year (tC/year). US per capita emissions exceed 5tC/year, and Japan and Western European nations emit 2 to 5 tC/year per capita. In the developing world, the per capita emissions are about 0.6tC/year, and more than 50 developing countries have emissions under 0.2tC/year. The global CO_2 emissions are far higher than what are understood as critical levels. Some note that:

> *... in order to prevent atmospheric GHG levels from exceeding twice the pre-industrial levels, average worldwide emissions must be stabilized at levels below 0.3 tC/year per capita for a future world population anticipated to stabilize near 10 billion people.* (Baer et al, 2000)

The debate on how to allocate CO_2 emission 'rights' (country versus per capita) began in the early 1990s. At the June 1991 meeting of the Intergovernmental Negotiating Committee (INC)[2] in Geneva, the Indian delegation argued for future convergence towards the same level of per capita CO_2 emissions. The reason for India and, also, China to raise the issue was to highlight the differences in responsibility of causing climate change. In this respect, China and India were relatively successful. Article 3 of the United Nations Framework Convention on Climate Change states:

> *The Parties should protect the climate system for the benefit of present and future generations of humankind on the basis of equity and in accordance with their common but differentiated responsibilities and respective capabilities. Accordingly, the developed country Parties should take the lead in combating climate change and the adverse effects thereof.*

The work towards the protocol that later came to be known as the Kyoto Protocol started when the 1995 Conference of the Parties (COP 1) in Berlin adopted the Berlin Mandate. This was a commitment to arrive at an agreement at COP 3 in 1997 that would include 'quantified limitation and reduction objectives' for Annex 1 parties (the industrial countries). Some time afterwards, the Ad Hoc Group on the Berlin Mandate (AGBM) was initiated to provide the venue for debating and negotiating the burden-sharing scheme that would become the foundation of the Kyoto Protocol. In a review of a range of proposals submitted to the AGBM process, Ringius et al (2002) note that proposals from France, Switzerland and the European Union (EU) built on the idea of a process towards future convergence of per capita emissions.

However, the issue has not been raised much in the COP meetings since the AGBM process. Instead, sovereign equality has been the basis for accounting for emissions. The total amount of emissions within territorial space is estimated and a 'cap' is decided upon. This is the architecture of the Kyoto Protocol, where all Annex 1 parties agreed to reduce emissions by at least 5 per cent from 1990 levels by 2008 to 2012. This kind of approach also 'freezes' time by deciding upon 1990 as the 'base year'. Since the 'cap' is based on past emissions, the Kyoto agreement rewards historically high emitters. It is the principle of sovereignty that lurks in the background as the implicit fairness principle. Current levels of emissions constitute a *status quo* right, and emissions should therefore be reduced proportionally across all countries in order to maintain relative emission levels between them (Ringius et al, 2002, p5).

In the climate negotiations, this is often called 'grandfathering', where allocation is done in proportion to a baseline of emissions. Certain territorial communities and their activities become naturalized through the reference to 'grandfathers' of the community. This is well in accordance with the idea of the international in that nations are intergenerational communities, bonding the present to the past and the future (Deudney, 1996). The idea of grandfathers is also seen, albeit somewhat diluted, in the principle of 'common but differentiated responsibilities'. It is furthermore a crucial principle in international fisheries management with reference to 'historic catch' (Ringius et al, 2002).

In some senses, the 'equal per capita' position poses a challenge to territoriality in that it relates individual human beings to the atmosphere as a global common. It establishes equal access and responsibilities for individuals on a new ground. However, individual CO_2 contributions are actually calculated on the basis of national emissions and then divided by the number of people living in the state. The 'per capita position' takes the emissions of the national community and divides by the population. Hence, the low Indian per capita level depends upon the large amount of poor Indians with very low emissions, while the Indian middle class consumes carbon on an Organisation for Economic Co-operation and Development (OECD) standard. This statistical construct, which Conca (2003) has called the 'myth of the average citizen', masks inequalities based on, for example, race, gender or ethnicity. The construction of an average carbon consumer provides a foundation for the state as a coherent and purposeful entity that can engage in negotiations to achieve justice for its citizens. In sum, the per capita position is also expressive of justice as a predominantly territorial discourse in the climate context. Worldwide consumers of carbon are much more territorialized than individualized.

CONCLUSIONS

Territoriality – that is, the idea that political authority is spatially organized and marked by clear boundaries – has a long lineage in political analysis. The linguist Paul Chilton notes that the elaboration of the sovereign state in Hobbes's Leviathan is cognitively linked to a 'container' image:

> *The political came to be imagined in spatialized terms and, specifically, through the spatial gestalt of the container which grounds the notions (and feelings) of identity and difference, of self and other, sovereign state and anarchic non-state, clearly and distinctly separated by a bounding limit.* (Chilton, 1996)

This chapter has illustrated how territoriality continues to shape the way in which the international community understands and manages the climate issue. Attempts to resist the territorial logic have been visible, for example, in Agarwal and Narain's well-known article 'Global warming in an unequal world: A case of environmental colonialism' (1991). They wrote in response to a report by the World Resources Institute (WRI) that had calculated the developing world share of the contributions to the accumulation of atmospheric CO_2 to be 48 per cent. Agarwal and Narain's calculations showed that developing countries were only responsible for 16 per cent. The rationale of their approach was:

> *No country can be blamed for the gases accumulating in the Earth's atmosphere until each country's share in the Earth's cleansing ability has been apportioned on a fair and equitable basis. Since most of the cleansing is done by the oceans and troposphere, the Earth has to be treated as a common heritage of mankind.* (Agarwal and Narain, 1991)

Agarwal and Narain proportionally distributed the world's sink capacity to each country in relation to its share of the world's population. Thus, 'India with 16 per cent of the world's population, gets 16 per cent of the Earth's natural air and ocean "sinks" for carbon dioxide and methane absorption' (Agarwal and Narain, 1991). Agarwal and Narain thus raise the question of to whom the world's sink capacity actually belongs and how it should be distributed. However, this has not been an issue in the climate negotiations. These negotiations seem, rather, to enhance rich countries' control over territories in both the North and the South.

Article 3.3 in the Kyoto Protocol mandates every country to report changes due to afforestation, deforestation and reforestation in domestic carbon pools over the commitment period (2008 to 2012). Furthermore, Article 3.4 of the

Kyoto Protocol gives countries the option to account for carbon stored in managed forests and agricultural lands since 1990 when meeting their commitments to limit greenhouse gas emissions. Although Sweden is a large country with vast forest areas, the Swedish government has so far resisted pursuing an active 'sink policy', and discussions on how to enhance Swedish carbon pools are only beginning to emerge among relevant public authorities. However, if Sweden does develop a more active sink policy – for example, through the use of various steering instruments – this will probably imply that another (potentially conflicting) dimension is added to discussions about the proper use and management of the Swedish landscape.

If this looks like a rather mild scenario within the territory of a rich Northern country, the sink issue within the Kyoto Protocol's Clean Development Mechanism is certainly raising far more difficult questions of power, authority and territory. Since these sinks are generated by tree plantation projects in the developing world (financed by Northern investors and lasting 20 or 30 years), the connection between a specific forestry activity, the resulting sink and the local territory becomes much tighter. Overall, this ability to control distant subordinate places more closely resembles imperial relations than international relations. In an imperial world, territories are not sovereign; but the logic of territoriality as rule over space still holds, albeit in a more complex form. While the climate negotiations under the UNFCCC have entered a rather technical and 'depoliticized' implementation phase, conflicting issues will surely surface again if concrete sink projects within the CDM become a popular method of complying with national CO_2 commitments.

The salience of territoriality in defining the contours of climate governance is also visible in the debate over the proper basis for counting emissions of carbon dioxide (country versus per capita). While this debate on how to allocate CO_2 emissions and emissions cuts is often understood as the end points of a broad spectrum, it is important to realize that the whole spectrum takes territorially based political communities for granted.

Almost 15 years ago, Hurrell and Kingsbury (1992) summed up the question that has driven so many of the debates within international environmental politics:

> Can a fragmented and often highly conflictual political system made up of 170 sovereign states and numerous other actors achieve the high (and historically unprecedented) levels of cooperation and policy coordination needed to manage environmental problems on a global scale?

The implicit assumption underneath Hurrell and Kingsbury's question is that environmental problems enter a world defined by sovereign territoriality, and that the management of our common global environment reflects this predicament. As argued in this chapter, however, it is not just a question of how to govern the climate in a bordered world because the governing of the climate actually borders *and* orders the world. It is nowadays conventional wisdom to argue for a changing role of the state among globalizing political patterns (see

150

for example, Scholte, 2001). It is also commonplace to argue that the landscape of international environmental politics is broader than the hegemony of liberal environmentalism might suggest (see, for example, Stevis, 2006). Despite this, territoriality seems to hold its grip on our political imagination, perhaps thereby limiting views of possible ways of coping with climate change.

NOTES

1   This section draws partly on Lövbrand and Stripple (2006).
2   By resolution 45/212 on 21 December 1990, the United Nations General Assembly launched negotiations on a climate change convention through establishing an Intergovernmental Negotiating Committee.

REFERENCES

Abbott, J. (1997) 'Export processing zones and the developing world', *Contemporary Review*, vol 270, pp232–238

Agarwal, A. and Narain, S. (1991) 'Global warming in an unequal world: A case of environmental colonialism', *Earth Islands Journal*, spring, pp39–40; reprinted as 'Global warming in an unequal world: A case of environmental colonialism', in Conca, K. and Dabelko, G. D. (eds) (1998) *Green Planet Blues*, Westview Press, Boulder, Colorado

Albert, M. (1999) 'On boundaries, territory and postmodernity: An international relations perspective', in Newman, D. (ed) *Boundaries, Territory and Postmodernity*, Frank Cass Publishers, London

Anderson, D., Grant, R. and Rolfe, C. (2001) *Taking Credit: Canada and the Role of Sinks in the International Climate Negotiations*, David Suzuki Foundation, Canada

Arrhenius, S. (1896) 'On the influence of carbonic acid in the air upon the temperature of the ground', *Philosophical Magazine*, vol 41, pp237–276

Baer, P., Harte, J., Haya, B., Holdren, J., Hultman, N. E., Kammen, D. M., Norgaard, R. B. and Raymond, L. (2000) 'Equity and greenhouse gas responsibility', *Science*, vol 289, p2287

Barkawi, T. and Laffey, M. (2002) 'Retrieving the imperial: Empire and international relations', *Millennium Journal of International Studies*, vol 31, pp109–127

Begg, K. G. (2002) 'Implementing the Kyoto Protocol on climate change: Environmental integrity, sinks and mechanisms', *Global Environmental Change*, vol 12, pp331–336

Benton, T. and Redclift, M. (eds) (1994) *Social Theory and the Global Environment*, Routledge, London and New York

Biersteker, T. J. (2002) 'State, sovereignty and territory', in Carlsnaes, W., Risse, T. and Simmons, B. A. (eds) *Handbook of International Relations*, Sage Publications, London

Biersteker, T. J. and Weber, C. (1996) *State Sovereignty as Social Construct*, Cambridge University Press, Cambridge and New York

Bolin, B. (1981) *Carbon Cycle Modelling*, New York, Chichester

Booth, K., Dunne, T. and Cox, M. (2000) 'How might we live? Global ethics in a new century – introduction', *Review of International Studies*, vol 26, pp1–28

Brown, A. D. (2001) 'The ethical dimensions of global environmental issues', *Daedalus*, vol 130, pp59–76

Bull, H. (1977) *The Anarchical Society: A Study of Order in World Politics*, Columbia University Press, New York

Caporaso, J. A. (2000) *Continuity and Change in the Westphalian Order*, Blackwell Publishers, Malden, MA

Chilton, P. (1996) 'The meaning of security', in Beer, F. A. and Hariman, R. (eds) *Post-Realism: The Rhetorical Turn in International Relations*, Michigan State University Press, East Lansing

Conca, K. (2003) 'Imagining the state', in Maniates, M. (ed) *Encountering Global Environmental Politics: Teaching, Learning and Empowering Knowledge*, Rowman and Littlefield Publishers Inc, Lanham

Conca, K. (2006) *Contentious Transnational Politics and Global Institution Building*, MIT Press, Cambridge, MA

Cox, M. (2004) 'Forum on the American Empire', *Review of International Studies*, vol 30, p583

Dalby, S. (2000) *Geopolitical Change and Contemporary Security Studies: Contextualizing the Human Security Agenda*, Working Paper no 30, April, Institute of International Relations, University of British Columbia, Vancouver, Canada, available at www.iir.ubc.ca/pdffiles/webwp30.pdf

Dalby, S. (2004) 'Ecological politics, violence, and the theme of Empire', *Global Environmental Politics*, vol 4, pp1–11

Dessai, S. (2001) 'Why did the Hague climate conference fail?', *Environmental Politics*, vol 10, pp139–144

Deudney, D. (1996) 'Ground identity: Nature, space and place in nationalism', in Lapid, Y. and Kratochwil, F. (eds) *The Return of Culture and Identity in IR Theory*, Lynne Rienner, Boulder, CO, and London, UK

Eckersley, R. (2004) *The Green State: Rethinking Democracy and Sovereignty*, MIT Press, Cambridge, MA

Fogel, C. (2004) 'The local, the global, and the Kyoto Protocol', in Jasanoff, S. and Martello, M. L. (eds) *Earthly Politics: Local and Global in Environmental Governance*, MIT Press, Cambridge, MA

Franke, M. F. N. (2000) 'Refusing an ethical approach to world politics in favour of political ethics', *European Journal of International Relations*, vol 6, pp307–333

Gossling, S. (2002) 'Global environmental consequences of tourism', *Global Environmental Change*, vol 12, pp283–302

Grubb, M., Vrolijk, C., Brack, D. and Energy and Environmental Programme (Royal Institute of International Affairs) (1999) *The Kyoto Protocol: A Guide and Assessment*, RIIA/Earthscan, London

Grubb, M. and Yamin, F. (2001) 'Climatic collapse at The Hague: What happened, why, and where do we go from here?', *International Affairs*, vol 77, pp261–276

Hardt, M. and Negri, A. (2000) *Empire*, Harvard University Press, Cambridge, MA

Houghton, R. A., Boone, R. D., Melillo, J. M., Palm, C. A., Woodwell, G. M. N., Myers, N., Moore, B. and Skole, D. L. (1985) 'Net flux of carbon dioxide from tropical forests in 1980', *Nature*, vol 316, pp617–620

Hurrell, A., and Kingsbury, B. (eds) (1992) *The International Politics of the Environment: Actors, Interests, and Institutions*, Oxford University Press, Oxford

Ikeme, J. (2003) 'Equity, environmental justice and sustainability: Incomplete approaches in climate change politics', *Global Environmental Change – Human and Policy Dimensions*, vol 13, pp195–206

Ikenberry, G. J. (2004) 'Illusions of empire: Defining the new American order', *Foreign Affairs*, vol 83, pp144–154

Jacoby, H. D., Prinn, R. G. and Schmalensee, R. (1998) 'Kyoto's unfinished business (on global warming)', *Foreign Affairs*, vol 77, pp54–66

Johnson, C. A. (2004) *The Sorrows of Empire: Militarism, Secrecy, and the End of the Republic*, Metropolitan Books, New York

Krasner, S. D. (1999) *Sovereignty: Organized Hypocrisy*, Princeton University Press, Princeton, NJ

Kuehls, T. (1996) *Beyond Sovereign Territory: The Space of Ecopolitics*, University of Minnesota Press, Minneapolis

Kuehls, T. (1998) 'Between sovereignty and the environment: An exploration of the discourse of government', in Litfin, K. (ed) *The Greening of Sovereignty in World Politics*, MIT Press, Cambridge, MA

Lafferreiere, E. (1994) 'Environmentalism and the global divide', *Environmental Politics*, vol 3, pp91–113

Lefebvre, H. (1991) *The Production of Space*, Blackwell Publishers, Oxford, UK, and Cambridge, MA

Lövbrand, E. and Stripple, J. (2006) 'The climate as political space: On the territorialization of the global carbon cycle', *Review of International Studies*, vol 32, pp217–235

Noble, I., and Scholes, R. J. (2001) 'Sinks and the Kyoto Protocol', *Climate Policy*, vol 1, pp5–25

Ó Tuathail, G. (1996) *Critical Geopolitics: The Politics of Writing Global Space*, University of Minnesota Press, Minneapolis

Ott, H. E. (2001) 'Climate change: An important foreign policy issue', *International Affairs*, vol 77, pp277–296

Paasi, A. (2003) 'Territory', in Agnew, J. A. (ed) *A Companion to Political Geography*, Blackwell Publishers, Oxford, UK, and Malden, MA

Paterson, M. (2001) 'Principles of justice in the context of climate change', in Luterbacher, U. and Sprinz, D. F. (eds) *International Relations and Global Climate Change*, MIT Press, Cambridge, MA

Ringius, L., Torvanger, A. and Underdal, A. (2002) 'Burden sharing and fairness principles in international climate policy', *International Environmental Agreements: Politics, Law and Economics*, vol 2, pp1–22

Rodhe, H., Charlson, R. and Crawford, E. (1997) 'Svante Arrhenius and the greenhouse effect', *Ambio*, vol 26, pp2–5

Rosa, L. P. and Munasinghe, M. (2003) *Ethics, Equity, and International Negotiations on Climate Change*, Edward Elgar, Cheltenham, UK, and Northampton, MA

Rowlands, I. H. (1997) 'International fairness and justice in addressing global climate change', *Environmental Politics*, vol 6, pp1–30

Rowlands, I. H. (2001) 'The Kyoto Protocol's "Clean Development Mechanism": A sustainability assessment', *Third World Quarterly*, vol 22, pp795–811

Sack, R. D. (1986) *Human Territoriality: Its Theory and History*, Cambridge University Press, Cambridge, UK

Scholte, J. A. (2001) 'The globalization of world politic', in Baylis J. and Smith, S. (eds) *The Globalization of World Politics: An Introduction to International Relations*, Oxford University Press, New York

Schultze, E.-D., Valentino, R. and Sanz, M. J. (2002) 'The long way from Kyoto to Marrakesh: Implications of the Kyoto Protocol negotiations for global ecology', *Global Change Biology*, vol 8, pp505–518

Sedjo, R., Marland, G. and Fruit, K. (2001) 'Accounting for sequestered carbon: The question of permanence', *Environmental Science and Policy*, vol 4, pp259–268

Shiva, V. (1994) 'Conflicts of global ecology: Environmental activism in a period of global reach', *Alternatives*, vol 19, pp195–207

Soja, E. W. (1989) *Postmodern Geographies: The Reassertion of Space in Critical Social Theory*, Verso, London and New York

Soroos, M. S. (1997) *The Endangered Atmosphere: Preserving a Global Commons*, University of South Carolina Press, Columbia, SC

Steinberg, P. E. (2001) *The Social Construction of the Ocean*, Cambridge University Press, Cambridge and New York

Stevis, D. (2006) 'The trajectory of the study of international environmental politics', in Betsill, M. M., Hochstetler, K. and Stevis, D. (eds) *Palgrave Advances in International Environmental Politics*, Macmillan, New York

Tonn, B. (2003) 'An equity first, risk-based framework for managing global climate change', *Global Environmental Change*, vol 13, pp295–306

Toth, F. L. (ed) (1999) *Fair Weather? Equity Concerns in Climate Change*, Earthscan, London

UNFCCC (United Nations Framework Convention on Climate Change) (1992) *United Nations Framework Convention on Climate Change*, www.unfccc.int

United Nations (1997) *Kyoto Protocol to the United Nations Framework Convention on Climate Change*, www.unfccc.int/essential_background/kyoto_Protocol/items/1678.php

Walker, R. B. J. (2002a) 'International/inequality', *International Studies Review*, vol 4, pp7–24

Walker, R. B. J. (2002b) 'They seek it here, they seek it there: Locating the political in Clayoquot Sound', in Magnusson, W. and Shaw, K. (eds) *A Political Space: Reading the Global through Clayoqout Sound*, University of Minnesota Press, Minneapolis and London

WCED (World Commission on Environment and Development) (1987) *Our Common Future*, Oxford University Press, Oxford

10

Multilevel Governance: A Solution to Climate Change Management?

Anders Biel and Lennart J. Lundqvist

CLIMATE CHANGE AS A SOCIAL DILEMMA

Ever since Hardin's classic 'The tragedy of the commons' (Hardin, 1968), there has been vital research on the commons and their regulation. Hardin presented a pessimistic view on the human ability to manage the commons, and advocated coercion. However, later experimental (for an early review, see Dawes, 1980) as well as empirical studies of local common-pool resource dilemmas (e.g. Ostrom, 1990; Ostrom et al, 2002) present a more positive view. People have been shown to voluntarily cooperate to manage the commons.

Admittedly, some commons are more easily managed than others. In their work on common-pool resource management, Ostrom et al (1994) make a distinction between stationary or spatially fixed resources and non-stationary resource units. While, for example, groundwater basins are stationary, many species of fish are not. They also make a division between available and unavailable storage, and whether stored units can be appropriated when needed. The combination of non-stationary and unavailable storage results in dilemmas where the costs of obtaining reliable information about the resource are high. Hence, such dilemmas are most difficult to balance. An attempt to solve a resource dilemma by Canadian fishermen was based on their catches staying on par with the birth rate of the fish in order to reach an optimal harvest level (Allen and McGlade, 1987). At the outset, catch rates decreased and the pool of fish was stabilized. Those were the days.

These physical qualities also have implications for designing effective institutions to govern the commons. In particular, rules of resource use based on quantity or space contribute to a successful management of common pool resources (CPRs) (Ostrom et al, 1994). Such rules are hard to establish when resources are not available in storage and are non-stationary. The size of the resource is uncertain and people face an appropriation problem (Gardner et al, 1990). Examples of local resources that have been studied are forests, groundwater systems, irrigation systems and inshore fish. Although information about the condition of the resource is far from perfect, appropriators do receive feedback about the effects of their actions. Depending upon the kind of resource, the size of the pool is more or less certain and environmental uncertainty is not complete.

Difficulties increase when we turn from local and regional to global CPRs (Biel, 2000). New obstacles relate both to the resource at hand and to social factors. Information about large-scale resources such as seawater, air and climate is far more difficult to collect. Feedback is often delayed and people face a social trap (Platt, 1973). Today, people see individual positive consequences from adopting a certain behaviour, but negative consequences from refraining, while the positive consequences for all from refraining appear in the future, as do the negative consequences from adopting the behaviour. Resource use results in immediate rewards but leads to long-term punishment.

Expressed differently, such decision situations have a deficient equilibrium. Given the delayed feedback, it may even be difficult to establish that negative consequences do follow from the original behaviour. Benefits and costs are detached. Moreover, they are differentially distributed across groups of people. Global common resources are also more difficult to monitor than local ones. Hence, environmental uncertainty increases, and so does uncertainty about the condition or size of the resource. When there is environmental uncertainty, people tend to overestimate the size of the resource and overuse it (see, for example, Gustafsson et al, 1999).

Some basic qualities that contribute to successful management are communication, trust and reciprocity. When people are able to communicate face to face, they can inform each other about the dilemma, make behavioural commitments and develop a group identity, and, as a result, breed trust and reciprocation. In local dilemmas, people can also monitor individual behaviour and develop and enforce sanctioning systems targeted at those who defect (see, for example, Rova, 2004). In complex dilemmas, people frequently act without explicit awareness of the dilemma that they are facing (Biel, 2000). Hence, face-to-face communication about the dilemma is rare; people act under anonymity and are less apt to monitor the behaviour of others. These are conditions where social norms will have a weak impact (Kerr, 1995).

Climate is a prime example of a collective resource exhibiting the combination of non-stationary and unavailable storage, where the costs of obtaining reliable information about the resource are high. Given that this global common-pool resource is extremely difficult to manage, it would seem a good idea to benefit from what research has taught us about the management of smaller commons and divide this global resource into smaller units, all in order to escape the 'big pool' illusion, which suggests that a resource may seem almost endless (Messick and McClelland, 1983). Experimental research shows that for various reasons, cooperation rates decrease as group size increases (e.g. Messick and Brewer, 1983). With smaller units, fewer people are responsible for the management of each unit and the problem of surveillance decreases. Furthermore, behavioural effects may become more visible (Olson, 1965; Ostrom, 1990). People act under less anonymity and social norms become more salient. Such a vision has been labelled bioregionalism (Sale, 1991). Empirical research shows its effectiveness in cases such as lobster fishing along the cost of Maine in the US (Acheson, 1987) and bleak-roe fishing in the Gulf of Bothnia (Rova, 2004).

NESTED DILEMMAS

Multilevel governance systems, paired with the idea of subsidiarity, partly work along these lines. Institutional arrangements are hierarchical and nested, and co-management exists between levels. An important proviso is that problems should be addressed on the most appropriate level. With respect to climate, the Kyoto Protocol was negotiated and signed at the international level. Co-management within the European Union (EU) comprises specific demands and commitments at the national level. Finally, regional and local levels have become involved in the implementation. Within and between all these levels, concordance ought to prevail concerning goals and appropriate policy measures in order to approach these goals.

Our analyses direct particular emphasis towards the local level. To what extent do they evidence and/or imply a successful implementation of a national strategy to manage the climate? Let it first be recognized that the investigated environmental policies to combat climate change relate to different levels, from the global to the local. In addition, each analysis incorporates policies that affect several levels or different target groups at the same level. It is not certain that these levels or user groups may recognize either the need of addressing climate change or the objectives of the climate strategy. Furthermore, other goals may interfere or be in conflict with the strategy. Provided that value priorities differ between levels and among groups, it may be difficult to reconcile them in a common governance of climate change.

Two of the chapters in this book concern policies at the international level. In Chapter 9, Stripple shows how the global issue of governing climate change is turned into a national problem through negotiations leading to the territorialization of climate space along traditional state borders. Here, national interests may come into conflict with the interests of the world at large. In Chapter 8, Gipperth analyses how commitments under the Kyoto Protocol are addressed within the EU and between member states in the EU. Also in this chapter, the interests of the larger community of Europe may be at odds with individual national concerns.

Most of the other chapters deal with policies that concern the national and local levels. In their analysis of the development of wind power in Sweden in Chapter 7, Söderholm et al show what happens with the Swedish policy for renewable and climate-friendly policy when it comes down on the municipal level. In Chapter 6, von Borgstede, Zannakis and Lundqvist investigate how professionals within and between different organizations at the local level respond to national policy measures, and in Chapter 5, Nilsson and Biel study how local decision-makers in the public sphere and in the market respond to national policy measures. Finally, in Chapter 4, Hammar and Jagers (as well as Bauhr in Chapter 3) analyse how national policies are received by representatives of the general public. Before some general characteristics of these conflicts are discussed, we present various specifics of each conflict in turn.

POTENTIAL CONFLICTS

While none of the chapters simultaneously investigate climate change from the global to the local level, together they show that the governance of climate change is truly nested in many layers. All of them, furthermore, observe the 'dilemma character' of environmental policies, where collective interests at a higher level may be in conflict with individual or collective interests at a lower level. In Chapter 9, Stripple argues that as national citizens, individuals are surrounded by different justice principles than they are as 'citizens of the world'. Since climate change does not coincide with national borders, the portioning of the climate resource along such borders in not beneficial in all respects when it comes to governing the climate as a commons.

When the EU ratified the Climate Convention, the union also declared that this was done jointly for the member states. A joint implementation implies an agreement on the extent of greenhouse gas emission reductions, as well as on how each country should meet this commitment. A burden-sharing agreement of an 8 per cent reduction in the EU was made in 1998, with rates varying from a reduction of 21 per cent in Germany and Denmark to an increase of 27 per cent in Portugal. While some member states have (over-)complied, others have not, evoking principles of distributive justice (see Chapter 8). In the best of worlds, the over-compliers would gladly compensate for those that do not in order to contribute in full to the common good. However, when national economic interests are at stake, we may see a call for compensation (and/or punishment of defectors).

Given global and national energy policies and a positive attitude among the general public, the prospect of wind power development in Sweden should be favourable. However, this has so far not been the case, and local governance seems to throw a spanner in the works. In Chapter 7, Söderholm et al point out two factors that primarily contribute to this slow progress. First, those who object to installations at the local level, mostly for aesthetic/environmental reasons, enjoy strong legal protection. Second, the municipality has a monopoly on physical planning within its territory and thus constitutes a formidable 'veto point'. As a result, national environmental and energy policy goals may not be promoted on the local level.

Additional examples of how local patterns can interfere with national goals and the implementation of policy instruments are provided by von Borgstede et al in Chapter 6. In order for a successful implementation of new policy instruments to take place, collaboration between organizations at the local level is an important prerequisite. However, the degree of involvement seems modest and less pronounced for organizations in the private sector. Organizational affiliation also interacts with the acceptance of new policy instruments. While both private and public organizations agreed that instruments such as information and subsidies were more acceptable than taxes and emissions trading, the degree of acceptance was higher within public than private organizations. Furthermore, acceptance co-varied with the position that respondents have within an organization. Those who, in their profession, worked with environmental issues were more prone to accept new climate policy measures than were groups of professionals who managed economic and planning matters, irrespective of whether they are in public or private organizations. This indicates that actors on the local level are guided by organizational cultures – that is, established organizational goals and values – in their response to new policy instruments. Since these cultures differ, unanimous support is difficult to establish. The importance of variation in organizational cultures is clearly shown by Nilsson and Biel (see Chapter 5). While public organizations were guided by environmental values and norms in their support of policy instruments, support in private organizations was, instead, dictated by internal goals linked to the benefit of the organization, rather than society at large. Nilsson and Biel also established that this view is linked to the professional role. Once professionals in private organizations respond to the same policy instruments in their 'private' life, their evaluations are in line with the general public (as well as with professionals in the public sector).

The public opinion on environmental policies is the topic for Bauhr (see Chapter 3) as well as for Hammar and Jagers (see Chapter 4). In Bauhr's case, the link to public opinion is indirect. She argues that to the extent that there is a consensus among experts on the causes and consequences of climate change, and how to deal with potential climate problems, trust in public institutions will increase. Hammar and Jagers take this finding a step further and show that, at least for increased carbon dioxide (CO_2) taxes, support increases with an increased trust in political institutions. It should be kept in mind, though, that the overall public support for increased CO_2 taxes is low. Rather, and in line with von Borgstede et al in Chapter 6, the support is much stronger for more lenient instruments, such as decreased taxes on fuels that do not affect the climate, expansion of public transport and more information on climate change effects related to traffic.

THE GENERAL PICTURE AND THE SWEDISH EXAMPLE

Policy goals and instruments are developed to protect our common good – the climate – and to avoid negative future consequences. These goals are developed on a global or European level, but are, to a large extent, instrumented and implemented on a national or local level. International future-oriented action towards climate change is first and foremost guided by environmental values and goals.

But once global or regional agreements and policies are passed on to the national and local levels for implementation, other values and goals make themselves felt. Despite adherence to the goal of environmental improvements, planning and decision-making are not always harmonized. Nations care about the financial costs and benefits related to CO_2 reductions and emissions trading. Local organizations are guided by organizational cultures and traditional frames for planning, where environmental values are not as strongly institutionalized as those cultures and frames. Individuals think about their own costs and convenience. Smaller efforts to combat climate change may be acceptable; but if people foresee larger sacrifices, their readiness for behavioural changes to support the collective good may not easily come forth.

How does this picture square with the three potential major shortcomings to behavioural changes that were proposed in Chapter 1: lack of knowledge, lack of motivation and lack of organizational and legal structures?[1] Lack of knowledge does not stand out as the prime obstacle. Rather, people are receptive to information, and research also shows a positive attitude among the general public towards behavioural changes in a more pro-environmental direction (see also, for example, Franzen, 2003).

However, knowledge and positive attitudes do not in themselves motivate people to make drastic changes in their life patterns. People may not only be guided by egoistic rather than collective or altruistic motives. Routine decision-making procedures may also preclude new motives. Although decision-making bodies at the international level sign agreements in order to promote environmental policies, the implementation of these policies at the local level is easier said than done. Decision-makers have established accepted frames of reference for evaluating policy options. To the extent that new criteria should influence the process, there must be a common understanding concerning how to apply and interpret the potential outcomes of these criteria. Such a consensus is not easily institutionalized. A strong motivation could facilitate an adoption. The fact that local organizations are prepared for smaller, but not for larger, behavioural changes indicates that although motivation is not lacking, it may not be strong enough.

Currently, adequate organizational and legal structures seem to be missing. Earlier research has shown the critical importance of community-based forms of organizing local resource use (e.g. Ostrom et al, 1994). However, once common pool resources are scaled up to larger units, whether they are national or global, other forms of institutional arrangements may be more suitable. Potential conflicts between decision-makers at different levels identified in the present work support this proposition. So do institutionalized 'glitches' between configurations of political/administrative authority and problem occurrence, as exemplified by the Swedish policies for wind power and for state support to local climate-related investments (see Chapter 7).

Hence, to portion the resource into smaller commons and to establish a multilevel governance system to cope with climate issues does not by itself guarantee that problems are solved. Nevertheless, such 'subsidiarity-driven' action may still be part of the solution.

What could be done – and is Sweden showing that it can be done?

Garret Hardin's essay on the 'Tragedy of the commons' appeared in *Science* in 1968. Thirty-five years later, *Science* honoured Hardin with a special issue, reviewing the research on local common-pool resource management. Summarizing general principles for governance of environmental resources, three of the leading researchers on governing the commons addressed global problems: Dietz, Ostrom and Stern (2003). Their general picture is that earlier attempts to govern resources by means of centralized command and control have failed. They thus point to the necessity to have institutional arrangements that are nested in layers. Their second principle concerns institutional variety; the nested layers should embrace a mixture of institutional types in order to employ a variety of decision rules and to induce compliance. The third principle is analytic deliberation among involved parties and the general public in order to improve information, increase trust and, hopefully, produce consensus on governance rules.

We see all three principles as relevant in order to increase the legitimacy of international and national climate strategies and the effectiveness of local implementation. In particular, they can contribute to reducing social uncertainty – that is, knowledge about how other parties will act. This results in increased trust. However, they do not directly address another kind of uncertainty that is crucial in common dilemmas: environmental uncertainty (Biel and Gärling, 1995). As already pointed out, this uncertainty concerns the condition or size of a resource, and when such uncertainty prevails, people tend to have an optimistic view of the condition or size of the resource and overuse it (Gustafsson et al, 1999). Better surveillance systems could contribute to reducing environmental uncertainty and, at the same time, to increasing trust in that resources are not overused.

As shown in this volume, Sweden's climate policy does seek to adhere to all of these three principles. What is obvious, however, is that the Swedish efforts so far have a mixed record with respect to both legitimacy and effectiveness. The nesting in layers tends to be rather 'path dependent'; authority is shared between the national government and the EU, on the one hand, and between the national government and the municipalities, on the other. There is so far not much evidence of a search for new layers, adopted for a more effective geographical handle on the climate change problem. And as our example of the Greater Gothenburg area shows, when actors do seek new ways of nesting, they fall back on existing structures, rather than contemplate new alternatives.

Furthermore, institutional variety may, indeed, seem to jeopardize both legitimacy and effectiveness in societal actions related to climate change. As evidenced particularly by the wind power case, a much needed expansion of this sustainable climate-friendly source of energy is delayed, if not derailed, as the large number of actors and interests with institutionalized rights to be involved in the decision-making process question the legitimacy of wind farm localization. Another example of how institutional variety may affect the legitimacy of climate policy concerns the EU emissions trading scheme. While laudable on grounds of effectiveness, the scheme (and its introduction of a whole new range of actors) practically 'lames' actors with a long-standing legitimacy in this field: the environmental courts and the local environmental bureaucracies.

When it comes to Sweden's record on analytical deliberation, let us first point to a general problem that is closely related to common dilemmas: the temporal and spatial separation of costs and benefits. People experience short-term benefits from activities that create costs in the environment that appear slowly and make themselves felt only in the distant future. Moreover, people who experience short-term benefits may be geographically separated from those who experience harm. Research has shown that people value immediate outcomes disproportionably higher than future outcomes. Hence, how to 'enlarge the shadow of the future' is an important task where policy instruments can play a role. In addition, research on social categorization suggests ways of building consensus. Recall that all studies in the programme identified potential conflicts between groups of decision-makers at different levels. Attached to their group identities, these groups come to the table with partly separate value structures and goals.

Social identity theory (e.g. Tajfel and Turner, 1986) and self-categorization theory (e.g. Turner et al, 1987) suggest that to the extent that people think of themselves as sharing a common identity rather than as separate individuals or groups, they become relatively more concerned about the welfare of the larger group. In the nested dilemmas that have been studied within our programme, all parties value the environment in the positive. However, established routines for decision-making may preclude environmental considerations. To the extent that a common identity on a higher level is emphasized, a goal transformation could take place and make environmental values come to the fore.

Does Sweden, with its strong tradition of open deliberation in order to reach as broad a consensus on political issues as possible on both national and local levels (see, for example, Lewin, 1998; Bäck, 2003), seem to be on the path to such a goal transformation? True enough, our studies imply that institutional variety may make such a development more difficult to achieve. But even if most studies in the programme identified potential conflicts between groups of decision-makers at different levels, it should be strongly noted that the local decision-makers interviewed all seem to accept the idea that climate change presents a common dilemma. One should also note that information fills an important function in developing common views, regardless of the contextual background. The semi-experimental study of opinions on carbon dioxide taxation implies that the more the revenues from such a tax are earmarked for longer-term environmental purposes, the easier it might be for the average taxpayer to accept such tax as legitimate. Furthermore, it would seem that the more problems and solutions to the common-pool resource dilemma of climate change are presented and deliberated in terms of willingness to accept, the more probable it is that the recognition of climate change as a common concern and responsibility might increase.

The Swedish example thus provides a two-pronged lesson. First, it shows that the all-encompassing issue of climate change brings forth new lines of conflicts and tensions within multilevel governance. Second, however, it implies that a political culture based on cooperation and consensus might be able to find ways of solving those conflicts and tensions. As one of our interviewees put it: 'To get to results, we must have information and communicate. Results come through dialogue and cooperation... This is the key' (von Borgstede and Lundqvist, 2007).

NOTE

1   Lack of adequate physical structures to promote pro-environmental behaviour is a problem that is not addressed in this volume.

REFERENCES

Acheson, J. M. (1987) 'The lobster fiefs, revisited: Economic and ecological effects of territoriality in the Maine lobster industry', in McCay, B. and Acheson, J. (eds) *The Question of the Commons*, University of Arizona Press, Tuscon, AZ

Allen, P. M. and McGlade, J. M. (1987) 'Modelling complex human systems: A fisheries example', *European Journal of Operations Research*, vol 30, pp147–167

Biel, A. (2000) 'Factors promoting co-operation in the laboratory, in common-pool resource dilemmas, and in large-scale dilemmas: Similarities and divergences', in van Vugt, M., Snyder, M., Tyler, T. and Biel, A. (eds) *Cooperation in Modern Society: Promoting the Welfare of Communities, States and Organizations*, Routledge, London

Biel, A. and Gärling, T. (1995) 'The role of uncertainty in resource dilemmas', *Journal of Environmental Psychology*, vol 15, pp221–233

Bäck, H. (2003) 'Party politics and the common good in Swedish local government', *Scandinavian Political Studies*, vol 26, pp93–123

Dawes, R. M. (1980) 'Social dilemmas', *Annual Review of Psychology*, vol 31, pp169–193

Dietz, T., Ostrom, E. and Stern, P. C. (2003) 'The struggle to govern the commons', *Science*, vol 302, pp1907–1912

Franzen, A. (2003) 'Environmental attitudes in international comparison: An analysis of the ISSP surveys 1993 and 2000', *Social Science Quarterly*, vol 84, pp297–308

Gardner, R. E., Ostrom, E. and Walker, J. M. (1990) 'The nature of common pool resource problems', *Rationality and Society*, vol 2, pp335–358

Gustafsson, M., Biel, A. and Gärling, T. (1999) 'Overharvesting resources of unknown size', *Acta Psychologica*, vol 103, pp47–64

Hardin, G. (1968) 'The tragedy of the commons', *Science*, vol 162, pp1243–1248

Kerr, N. L. (1995) 'Norms in social dilemmas', in Schroeder, D. (ed) *Social Dilemmas: Theoretical Issues and Findings*, Pergamon Press, New York

Lewin, L. (1998) 'Majoritarian and consensus democracy: The Swedish experience', *Scandinavian Political Studies*, vol 21, pp195–206

Messick, D. M. and Brewer, M. B. (1983) 'Solving social dilemmas: A review', *Review of Personality and Social Psychology*, vol 4, pp11–44

Messick, D. M. and McClelland, C. L. (1983) 'Social traps and temporal traps', *Personality and Social Psychology Bulletin*, vol 9, pp105–110

Olson, M. (1965) *The Logic of Collective Action*, Harvard University Press, Cambridge, MA

Ostrom, E. (1990) *Governing the Commons: The Evolution of Institutions for Collective Action*, Cambridge University Press, New York

Ostrom, E., Dietz, T., Dolšak, N., Stern, P. C., Stonich, S. and Weber, E. U. (2002) *The Drama of the Commons*, National Academy of Sciences Press, Washington, DC

Ostrom, E., Gardner, R. and Walker, J. (1994) *Rules, Games, and Common-Pool Resources*, University of Michigan Press, Ann Arbor, MI

Platt, J. (1973) 'Social traps', *American Psychologist*, vol 28, pp641–651

Rova, C. (2004) *Flipping the Pyramid: Lessons from Converting Top-Down Management of Bleak-Roe Fishing*, Luleå University of Technology, Luleå

Sale, K. (1991) *Dwellers in the Land: The Bioregional Vision*, 2nd edition, New Society Publishers, Philadelphia

Tajfel, H. and Turner, J. C. (1986) 'The social identity theory of intergroup behaviour', in Worchel, S. and Austin, W. G. (eds) *Psychology of Intergroup Relations*, 2nd edition, Nelson-Hall, Chicago

Turner, J. C., Hogg, M. A., Oakes, P. J., Reicher, S. D. and Wetherell, M. (eds) (1987) *Rediscovering the Social Group: A Self-Categorization Theory*, Blackwell, Oxford

von Borgstede, C. and Lundqvist, L. J. (2007) 'Whose responsibility? Swedish local decision-makers and the scale of climate change abatement', *Urban Affairs Review* (forthcoming in 2007)

Index